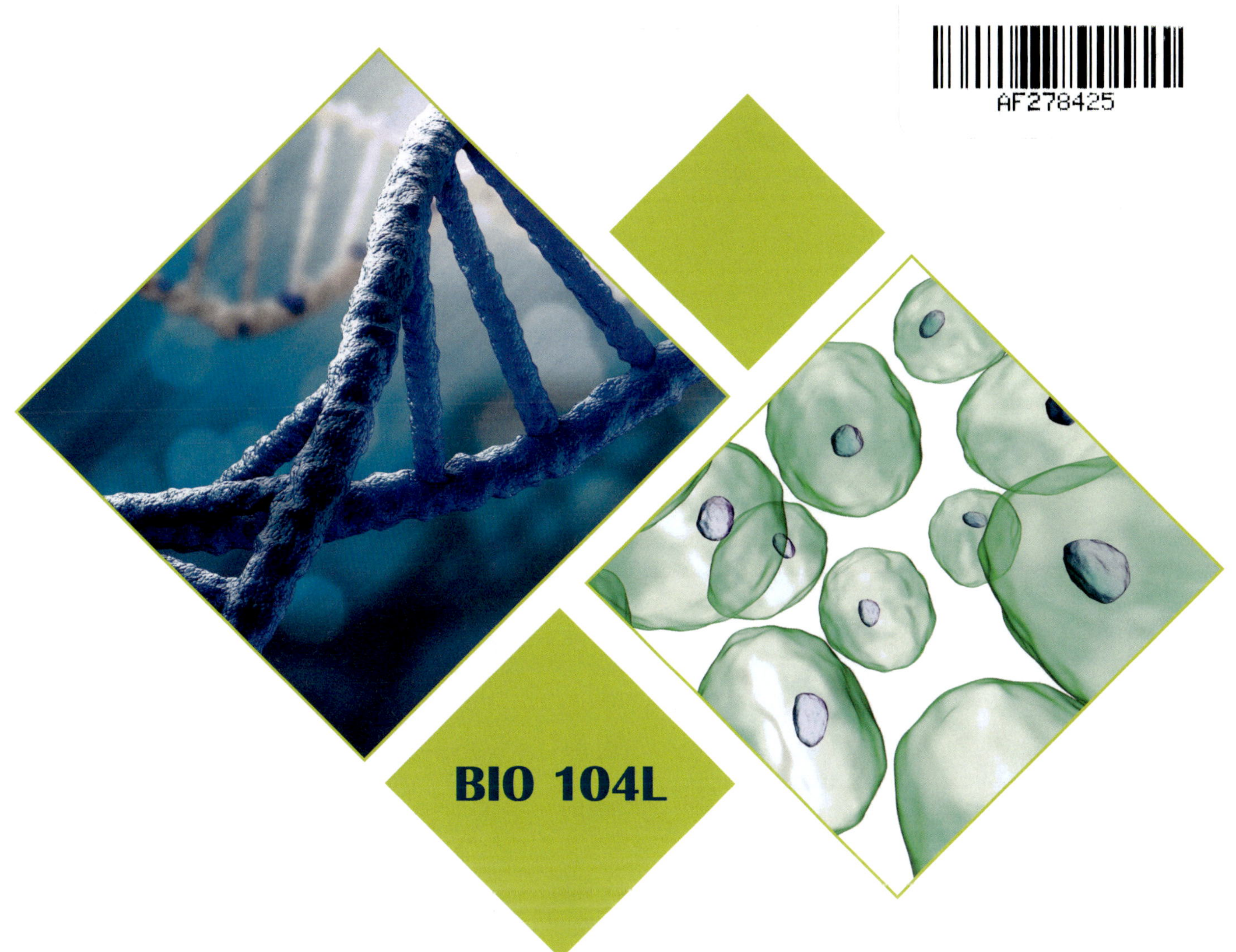

THE CELL & DNA LABORATORY

Cuyahoga Community College
Joseph Koch

The Cell & DNA Laboratory

BIO 104L
Cuyahoga Community College
Joseph Koch

Printed in the United States of America
10 9 8 7 6 5 4
ISBN: 978-1-61740-546-4

Van-Griner Publishing
Cincinnati, Ohio
www.van-griner.com

CEO: Mike Griner
President: Dreis Van Landuyt
Project Manager: Brenda Schwieterman
Customer Care Lead: Julie Reichert

Koch 546-4 Su18
196421-319062
Copyright © 2019

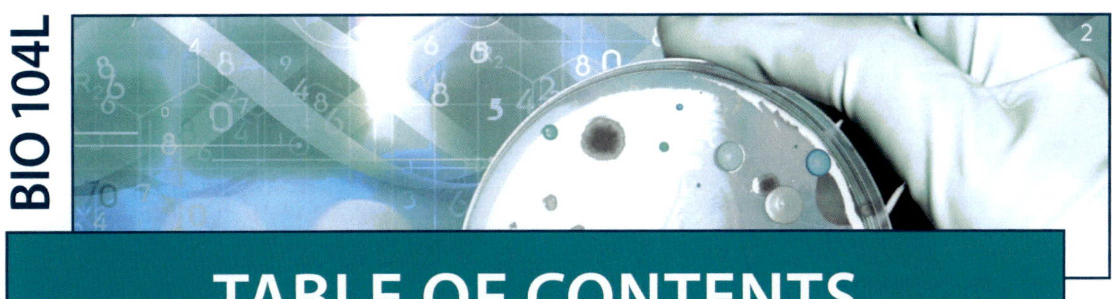

TABLE OF CONTENTS

LABORATORIES

APPENDICES

THE PROCESS OF SCIENCE

OBJECTIVES

By the end of this lab exercise, students should be able to

- distinguish between discovery-based science and hypothesis-based science;
- list and describe the major steps of the scientific method and apply them to example scenarios;
- explain how variables influence experiments and distinguish between dependent and independent variables;
- record observations of live animals;
- conduct a simple experiment and draw conclusions; and
- define **hypothesis, variable, independent variable, dependent variable,** and **control trial.**

INTRODUCTION

When students hear the term "science," they often picture people in white coats mixing chemicals in a lab. While there are plenty of scientists who work in this type of environment, science is a concept that applies to everyone. At its most fundamental level, science is a way of gathering knowledge. No matter what you study or do for a living, there will come a time when you need to gather knowledge about the world around you. In this respect, it is important that we understand the different ways in which scientific investigations can be carried out.

Scientific investigations can differ depending on the use of *discovery-based science* and *hypothesis-based science.* Discovery-based science typically involves observations and analyses that result in descriptive data. Think of this type of science as answering "who," "what," and "when" questions. Hypothesis-based science involves forming a **hypothesis** (a suggested explanation for a specific problem), which can be tested by carrying out an experiment. Hypothesis-based science can be thought of as answering "how" and "why" questions. In many cases, a specific scientific investigation may involve both basic types of sciences. Such investigations involve proper use of the Scientific Method, which you will explore in Procedure 1.1.

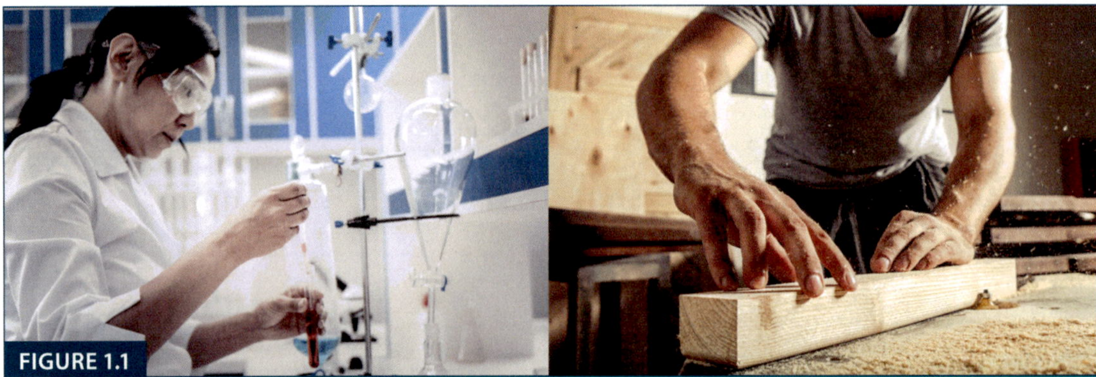

FIGURE 1.1

Scientific Investigations. While many scientific investigations are conducted in laboratories, others are routinely conducted in the field, at job sites, or at home. A carpenter can use the process of scientific inquiry to determine which type of wood is best for a project.

PROCEDURE 1.1: APPLYING THE SCIENTIFIC METHOD

Imagine that you are taking notes in class, when all of the sudden you notice the person next to you has had the misfortune of having their pen leak all over their shirt. Knowing that ink stains can be very difficult to remove, you decide to be proactive and try to find a way to remove ink from clothing in the event that it ever happens to you. While at a local department store, you notice three different detergents that claim to remove "tough stains." You set out to determine which, if any, can remove pen ink from a shirt.

The first step in the scientific method involves *making an observation* about the world around you that raises a specific question. This serves as a preliminary step because you aim to design an experiment that may help answer the question. In the example above, what observation was made, and what question would this observation raise?

The second step of the scientific method involves *forming a hypothesis,* which can be thought of as a potential answer to a question that has been raised. Using the example above, what hypothesis could you generate regarding the observation and question raised from the previous step?

After forming a hypothesis, the next step in the scientific method is to test the hypothesis by *designing and conducting an experiment.* This step is crucial because poorly-designed experiments can lead to misleading results and conclusions. Before we attempt to design an experiment using the example above, consider the following sample experiment:

To determine which detergent works best at removing ink stains, you test all three in succession. For the first detergent, you spill some ink from a red pen on a pair of jeans and wash them using the detergent. For the second detergent, you find a second pen (a black pen) and use it to stain a white t-shirt. You wash the shirt with the detergent. For the third detergent, you break open your last pen (a blue pen), and use it to stain a bed sheet before washing it. After testing all three detergents, you compare the stains to determine which detergent works best.

Thinking Critically

Discuss with your lab partners and try to come up with three problems with the experimental design described above.

When designing an experiment, one must always be aware of the **variables** that can impact the results. A variable is anything that can change during an experiment. Because variables can influence the experimental outcome, it is important to try and keep them constant between experimental trials. In the sample experiment above, the type of clothing used and color of ink are examples of a variables that should be kept constant in each trial.

There are two variables in an experiment that are *expected* to vary in each trial. The **independent variable** in an experiment is that which is varied in each trial, and is often thought of as the "cause" or the "treatment." For example, if a dietician were testing the effects of weight loss supplements, the type of supplement being used would be the independent variable. The **dependent variable** is the effect being measured in the experiment. In an experiment testing weight loss supplements, the dependent variable would be the amount of weight lost by the individuals in each trial.

What would be the independent and dependent variables of the detergent experiment?

Now that you have identified what ***not*** to do when designing an experiment, work with your lab partners to create an experimental design that could better test the detergents and their ability to remove ink. When you are finished, be sure to show your experiment to your instructor.

In many cases, the independent variable(s) being tested may impact the dependent variable significantly, but it can be difficult to draw conclusions without including a trial that produces a known outcome. In a **control trial,** the treatment being tested will produce a known result, which will create a standard of comparison for the other trials. Water is often used as a substitute for a reactant in an experiment, as it usually does not react strongly. In the detergent example, one trial might include water instead of detergent, which would demonstrate the outcome of using no cleaning agent at all on the stain.

PROCEDURE 1.2: MAKING OBSERVATIONS AND COLLECTING DATA

Recall that discovery-based science involves making basic observations and collecting data. Sometimes, before a hypothesis-based experiment can be conducted, we need to know more about the nature of what we are studying. It is from these basic observations that important questions might arise. In this procedure, you will make anatomical and behavioral observations of mealworms using a dissecting microscope (your instructor will show you how to set up and use the microscope). Mealworms may resemble worms in their larval state, but they are actually a type of beetle. Like all beetles, their adult state looks entirely different from their larval state (see Figure 1.2). Your instructor will provide you with a larval and adult mealworm to observe under your microscope. Record observations pertaining to a larval mealworm and adult mealworm in Table 1.1.

FIGURE 1.2

Larval and Adult Forms of a Mealworm Beetle, *Tenebrio molitor*

Observations of Mealworms

TABLE 1.1		
OBSERVATION	**LARVAL MEALWORM**	**ADULT MEALWORM**
Appearance/ Texture of Skin		
Locomotion		
Mouthparts		
Number of Legs		
Number of Body Segments		
Wings Present/ Absent		
Other Distinguishing Characteristics		

After collecting your observations, compare your data with another lab group. How did their observations compare to your own?

Do you think it could be beneficial for multiple scientists to conduct observational investigations on the same type of organism? Briefly explain why/why not.

PROCEDURE 1.3: CONDUCTING AN EXPERIMENT

In Procedure 1.1, you were given the opportunity to design an experiment based on an independent variable that was already given to you. You will now design and conduct an experiment that tests how reaction time can be influenced by a variable of your choosing. To do this, at least one member of your group must download a free app that measures reaction time. The app will ask you to point at a spot on your phone with your finger, and will register the time it took for you to complete this action. The more time that passes, the longer (slower) your reaction time. Your instructor will provide you with the name of the app.

It is up to your group to decide what independent variable you would like to test in your experiment. Think about things that might hinder or enhance your ability to react that may be worth testing. For example, if you wanted to test the effects of some form of distraction on reaction time, you could have certain individuals try having a conversation while performing the reaction test. Once you have determined what variable you want to test in your experiment, be sure to tell your instructor before beginning to design the experiment.

Use the space below to briefly outline your experiment before you carry it out. Be sure to include your **independent variable, the dependent variable,** and how you will carry out the experiment using different trials to test the independent variable. Include a **hypothesis** about how you think your independent variable will affect the dependent variable.

As you conduct your experiment, create a data table in the space below that documents your results. Your table should include reaction time data for each trial. If each person takes part in multiple trials, you should include their average reaction time score. Since you are testing the effects of some variable on reaction time, be sure to include the category that each person would fit into with regard to the independent variable being tested. For example, if you were testing the effects of distraction on reaction time, you would indicate if each trial run was from a distracted or not distracted person.

Measuring The Effects of
on Reaction Time

TABLE 1.2

Once you have recorded your data, complete the follow-up questions:

1. Was your initial hypothesis supported by your data? _______________________

2. What conclusions, if any, can you draw from your data?

3. What factors may have influenced the outcome of your experiment aside from the variable that you were testing?

2

TAKING MEASUREMENTS IN BIOLOGY

OBJECTIVES

By the end of this lab exercise, students should be able to

- use laboratory equipment to take measurements of length, mass, volume, and temperature;
- convert between units in the metric system; and
- convert between Fahrenheit and Celsius temperature readings.

INTRODUCTION

When scientists conduct experiments and gather data, it is important that they take precise measurements. It is therefore necessary that scientists become familiar with the devices used to take these measurements. In the labs to come, we will be measuring and collecting data pertaining to length, mass, volume, and temperature. Before we can begin to take these measurements, we must learn how to use the equipment that we have available to us in lab. Regardless of the type of equipment used, scientists always measure and record data using the *metric system*. Review the information in Table 2.1 and Table 2.2, which display divisions and multiples of metric units and their common prefixes. Notice that the metric system is based on units of *ten*. This makes converting from one unit to another relatively simple.

Divisions of Metric Units and Their Prefixes

TABLE 2.1

PREFIX	SYMBOL	DIVISION OF METRIC UNIT
Deci	d	0.1 (10^{-1})
Centi	c	0.01 (10^{-2})
Milli	m	0.001 (10^{-3})
Micro	μ	0.000001 (10^{-6})
Nano	n	0.000000001 (10^{-9})

Multiples of Metric Units and Their Prefixes

TABLE 2.2

PREFIX	SYMBOL	MULTIPLE OF METRIC UNIT
Deka	da	10 (10^{1})
Hecto	h	100 (10^{2})
Kilo	k	1,000 (10^{3})
Mega	M	1,000,000 (10^{6})
Giga	G	1,000,000,000 (10^{9})

To convert from *centimeters* (cm) to *meters* (m), you would thus multiply by 0.01, and to convert from *millimeters* (mm) to *meters* (m), you would multiply by 0.001. If you were to convert from centimeters to millimeters, you would first determine how many millimeters are in one centimeter (10). For example, to convert **45 cm** to **mm,** you would use the following equation:

$$45 \text{ cm} \times \frac{10 \text{ mm}}{\text{cm}} = 450 \text{ mm}$$

Notice from the equation above that the units you are converting *from* cancel out, leaving you with the units that you are converting *to* in your answer. Always remember that converting to a *smaller* unit will give you a *larger* number (i.e., 45 to 450). If you convert to a larger unit, your value will *decrease*.

If you can remember the common metric prefixes in the correct order, making conversions can be simple, as you only have to know how to move the decimal point when converting. It may be helpful to think of the units (in descending order) as a flight of stairs, where the decimal point moves up or down the stairs (left or right) depending on the units being converted (see Figure 2.1). For example, to convert **17.5 milligrams (mg)** to **decigrams (dg),** you would move the decimal point in your original number (17.5) *two spaces to the left* (two "steps" up from milligram) to get **0.175 dg.** If you were to convert a whole number to a different unit, the decimal point would begin after the number being converted. For example, to convert **35 decimeters (dm)** to **millimeters (mm),** you would move the decimal point *two spaces to the right* to get **3,500 mm.** When converting this way, remember to add additional "steps" when converting to/from Giga, Mega, Micro, and Nano. For example, to convert 41.8 milligrams (mg) to **micrograms (µg),** you would move the decimal point *three spaces to the right* to get **41,800 µg.**

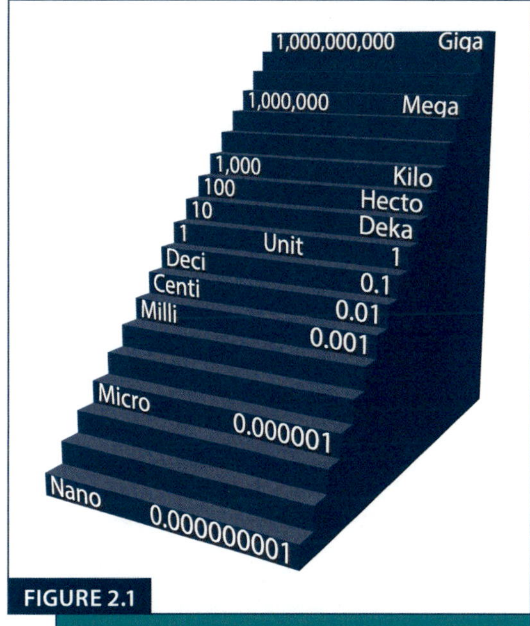

FIGURE 2.1

The Metric Staircase. As you move up the staircase, the decimal moves to the left, and as you move down, the decimal moves to the right.

PROCEDURE 2.1: MAKING METRIC CONVERSIONS

Practice converting metric units by making the following conversions:

1. 1 meter (m) = _____ centimeters (cm)

2. 5.7 meters (m) = _____ centimeters (cm)

3. 7.9 dekameters (dam) = _____ hectometers (hm)

4. 17.2 liters (L) = _____ milliliters (mL)

5. 6.5 centiliters (cL) = _____ liters (L)

6. 15 grams (g) = _____ kilograms (kg)

7. 13.9 milligrams (mg) = _____ micrograms (µg)

Thinking Critically

You may be used to taking measurements using the "standard" or "customary" system commonly used in the United States, which uses units like *inches* for length and *pounds* for mass. Scientists, even those in the United States, always use the metric system when conducting experiments. Discuss with your lab group members and come up with an explanation as to why it is important for scientists to use the same measurement system.

Now that you are familiar with the metric system, we can begin to take measurements. The following procedures require you to use specific lab equipment that you may have not used before. If you are unfamiliar with a specific piece of equipment, be sure to take extra practice before moving on to another procedure.

PROCEDURE 2.2: TAKING METRIC MEASUREMENTS OF LENGTH

Biologists routinely take length measurements when determining growth rates, making anatomical comparisons, and when tracking animal behavior. In our future labs, we will be measuring reaction rates in test tubes using rulers. The basic metric unit of length is the **meter (m).** To practice measuring length, use the meter sticks and rulers at your table to take the following measurements:

1. Length of this page in centimeters (cm): ____
2. Width of a pencil eraser in millimeters (mm): ____
3. Your height in centimeters (cm): ____
4. Width of this page in centimeters (cm): ____
5. Length of your cell phone in centimeters (cm): ____
6. *Area* of this page (***Hint:*** area = length × width): ____

Measure the diameter of the circle to the right using a ruler marked with centimeters (cm) and millimeters (mm). Take your measurement in millimeters.

Diameter of the circle: ____ mm

Now convert your measurement to micrometers (μm).

Diameter of the circle: ____ μm

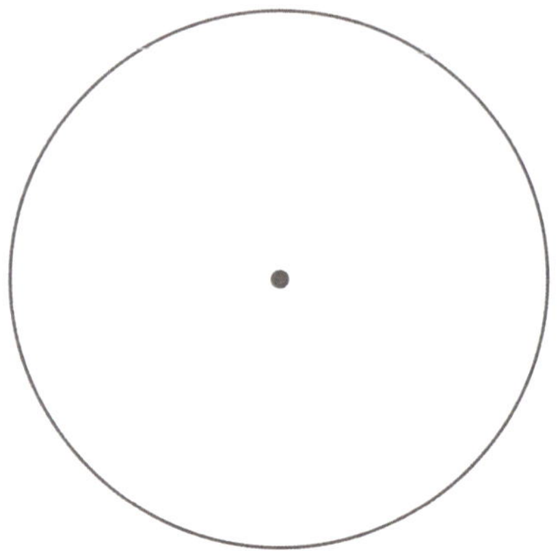

PROCEDURE 2.3: TAKING METRIC MEASUREMENTS OF MASS

Mass is a measurement of the amount of matter in a substance. Biologists often use electronic balances when measuring mass in experimental investigations (see Figure 2.2). Given that scales and balances can also be used to measure weight, students often confuse the two. Weight measures the force of gravity on an object, which can differ depending on environment. An object on the moon, for example, would weigh less than it would on Earth, but would have the same mass. Electronic balances do not measure mass directly, but they are calibrated to convert the weight of an object to a mass reading due to the relationship between mass and weight on earth.

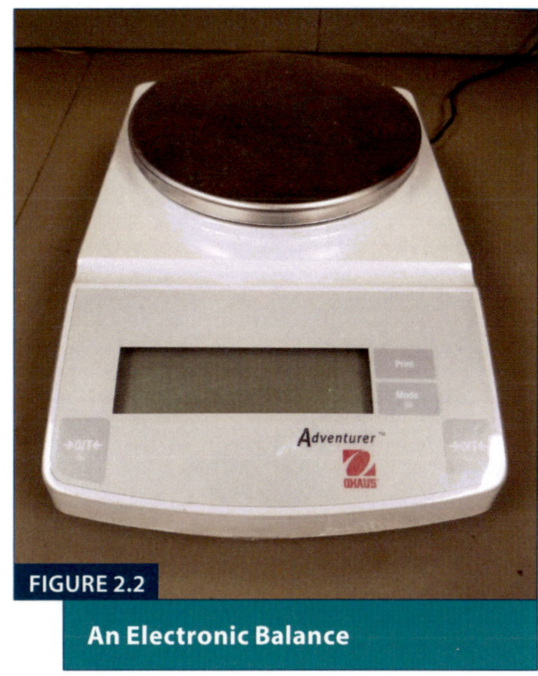

FIGURE 2.2

An Electronic Balance

Practice measuring mass using the electronic balances in the lab room. When using a balance, first make sure that the surface of the balance is clean. If there is visible debris on the surface of the balance, gently wipe it clean using a disposable wipe or paper towel. Before taking any measurement, be sure to *zero* the balance by pressing the "on" button, which is often also labeled "zero." Then place your item on the balance and record the measurement in grams (g). Record the masses of the following items:

1. A small wooden block: ___ g
2. A paper clip: ___ g
3. A cell phone: ___ g
4. A piece of cork: ___ g
5. A 100-mL beaker: ___ g

Suppose you wanted to measure the mass of 100 mL of water (see Figure 2.3). Discuss with your group members as to how you would do this, and record the mass that you measure below.

Mass of 100 mL of water: ___ g

PROCEDURE 2.4: TAKING METRIC MEASUREMENTS OF VOLUME

When biologists conduct experiments involving specific quantities of liquid, they do not normally use mass as their means of measurement because measuring the mass of a liquid is difficult without having to take into account the mass of the container holding the liquid (as you demonstrated in Procedure 2.3). Instead, volume is typically used as a means of measuring quantities of fluid. Volume is a measurement of the amount of space that an object occupies. The metric unit for volume is the *liter*. In this procedure, you will practice measuring the volume of liquids and solids using different instruments. You will also learn how to use a *serological pipette* to add or subtract specific volumes of fluid.

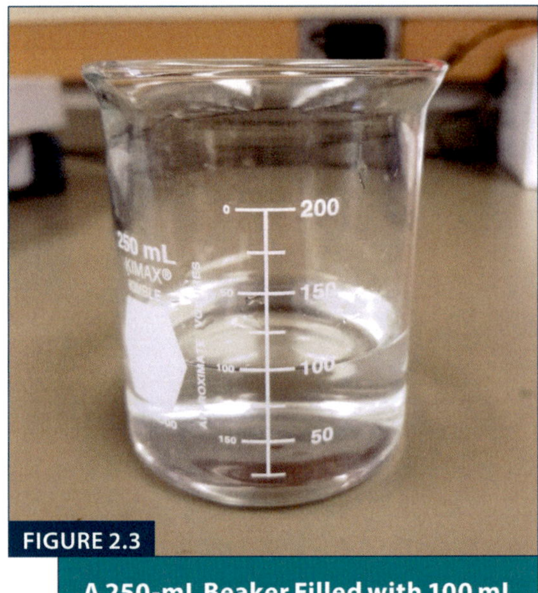

FIGURE 2.3

A 250-mL Beaker Filled with 100 mL of Water

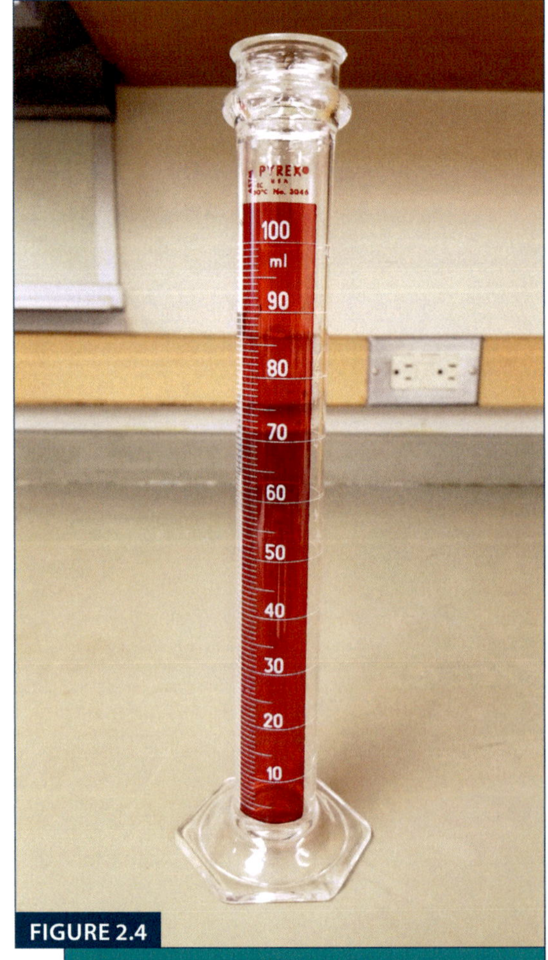

FIGURE 2.4

Graduated Cylinder

One of the most common pieces of lab equipment used to measure volume is the graduated cylinder (see Figure 2.4). As the name implies, graduated cylinders are *graduated,* which means they are marked with horizontal lines representing specific units (milliliters) in a scale. Some common graduated cylinders can measure up to 20 mL, and others measure up to 50 mL. To measure a specific volume of fluid using a graduated cylinder, simply pour the fluid into the cylinder until the fluid level reaches the desired volume line. It is very important to remember, however, that water molecules will adhere to glass, which will cause the fluid level to curve slightly up the sides of the cylinder to form a *meniscus.* When measuring precise volumes using a graduated cylinder, always read from the **bottom** of the meniscus (see Figure 2.5).

Discuss with your lab group members as to how you would determine the total volume of water (in mL) that a paper cup can hold using a graduated cylinder. Record your measurement below.

FIGURE 2.5

A Meniscus in a Test Tube

Total volume of water held by paper cup: ____ mL

Now discuss with your lab group members as to how, using a graduated cylinder, you would determine the volume of the piece of cork used in Procedure 2.3. To be sure that your method is accurate, inform your instructor as to how you plan to measure the volume of the piece of cork.

Total volume of piece of cork: ____ mL

Perform the same test to measure the volume of the small wooden block used in Procedure 2.3. Record the volume of this object below.

Total volume of wooden block: ____ mL

Thinking Critically

When performing a *water displacement* test, objects are submerged in water, typically in a graduated cylinder, to cause the water level to rise to a measureable level. When performing this test with an object that floats in water (like a cork), it is important to make sure that the entire object is submerged. Describe why a floating object must be fully submerged to get an accurate volume measurement. If you were pushing the object into the water with your finger, do you think pushing your finger below the surface of the water would affect the measurement? Explain your reasoning below.

PROCEDURE 2.5: CALCULATING THE DENSITY OF AN OBJECT

Density is a measurement of the compactness of a substance. The wooden block and cork from the previous experiment differed in their buoyancy, which required you to take an extra step when measuring the volume of the cork. Density, as you might expect, determines the buoyancy of an object in water, and is dependant upon the object's mass and volume.

Thinking Critically

If you were having a snowball fight with friends, you would probably want to make a snowball that could be thrown from a good distance without falling apart in your hands (or in the air). Assuming you had a small pile of snow of a certain mass, how would you make your snowball in order to make it ideal for throwing? In doing so, what have you done to the *density* of the pile of snow? What would happen to the *volume* of the pile of snow as you increase its density?

To measure the density of an object, divide its mass by its volume. If the mass and volume were taken in grams (g) and milliliters (mL) respectively, the calculated density can be expressed as grams per milliliter (g/mL).

$$d = \frac{M}{V}$$

You can thus calculate the density of the objects used in Procedures 2.3 and 2.4. Record the densities of the substances below.

1. Density of piece of cork: ____ g/mL

2. Density of wooden block: ____ g/mL

3. Density of water: ____ g/mL

Given your data, what can you conclude about the density of an object (relative to water) and its ability to float on water?

PROCEDURE 2.6: USING A SEROLOGICAL PIPETTE

Serological pipettes are routinely used by biologists to measure a specific volume of fluid that must be transferred to another container. Serological pipettes have a handle and a tip, which can be fastened together (see Figure 2.6).

When attaching the serological pipette tip to the handle, make sure that the tip is firmly in place so that air does not escape when drawing fluid into the tip. Notice that there is a scale of lines and numbers on the tip, similar to a graduated cylinder. Your instructor will demonstrate how to draw fluid to specific volume levels using a serological pipette. Practice using a serological pipette by transferring water from a beaker into a graduated cylinder.

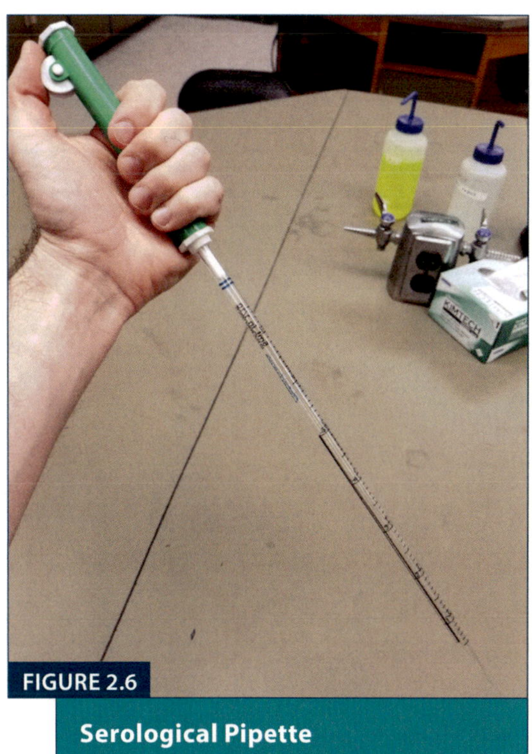

FIGURE 2.6

Serological Pipette

PROCEDURE 2.7: TAKING TEMPERATURE MEASUREMENTS USING A THERMOMETER

In biology, temperature is often a variable that can influence the rate at which a specific reaction occurs. In the labs to come, you will determine the effects of temperature changes on specific metabolic reactions. Here we will practice taking temperature readings and making conversions between two common temperature scales.

Pick up the thermometer in the bin at your table and notice the level of the red fluid within. Notice that your thermometer may have two different numerical scales (C and F). Figure 2.7 displays a thermometer with both scales. In the United States, temperatures (such as those associated with weather) are often described using the Fahrenheit scale. In this course, we will be measuring temperatures using the Celsius scale. In the Celsius scale, water freezes at 0 °C and boils at 100 °C. There are thermometers placed in specific environments around the lab room. Record the temperature (in °C) of each environment below by reading the fluid level in the thermometer.

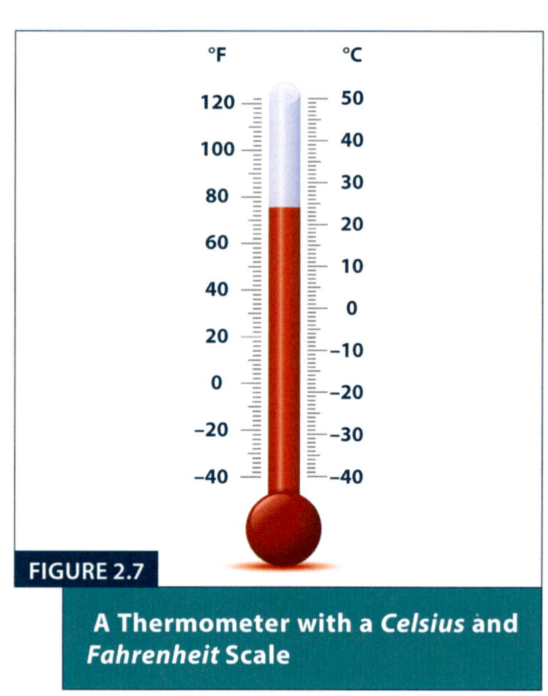

FIGURE 2.7

A Thermometer with a *Celsius* and *Fahrenheit* Scale

1. Room temperature (thermometer at your table): ___ °C

2. Refrigerator: ___ °C

3. Warm water incubator: ___ °C

4. Ice bath: ___ °C

If you had a thermometer that only had a Fahrenheit scale, you might need to convert your temperature reading to Celsius when conducting an experiment. To convert from Fahrenheit to Celsius, use the following formula:

$$°C = \frac{(°F - 32)}{1.8}$$

Using the Fahrenheit scale, record the temperature of the room in Fahrenheit (°F).

Room temperature in Fahrenheit: _____ °F

Now convert this value using the formula above to get the temperature of the room in °C. Compare your answer to the value you recorded when you took the measurement using the Celsius scale.

Room temperature in Celsius: _____ °C

If you needed to convert a temperature reading from Celsius to Fahrenheit, use the following formula:

$$°F = (1.8)(°C) + 32$$

Practice converting some of your previous Celsius temperature readings to the Fahrenheit scale using the equation above.

3

MICROSCOPY

OBJECTIVES

By the end of this lab exercise, students should be able to

- identify and describe the important parts of a compound light microscope;
- use a compound light microscope to view slide samples at different magnifications;
- practice microscope safety and adhere to the guidelines of proper microscope use;
- use a microscope to measure field of view;
- calculate the total magnification of a compound light microscope; and
- prepare and view a wet-mount slide.

INTRODUCTION

When biologists make use of the scientific method to answer questions about biology, they often have to collect data and make observations based on things that are too small to see with the naked eye. To truly understand living things, biologists must study the cells that make them up. The vast majority of cells are too small to see without the help of a **microscope,** which is an instrument that magnifies an object. Red blood cells, for example, are so small that you could fit roughly 250 of them on the head of a pin.

In this lab, we will be using a *compound light microscope.* The "compound" designation refers to the fact that the microscopes have two sets of lenses that work to magnify the specimen being viewed. The "light" designation refers to the fact that light from a bulb at the bottom of the microscope shines through the specimen and is bent by the lenses to provide an image to the viewer. The following procedures will help you learn how to properly use a compound light microscope.

PROCEDURE 3.1: LEARNING THE PARTS OF A COMPOUND LIGHT MICROSCOPE

Your instructor will now go over the important parts of your compound light microscope. As you cover each part, be sure to identify the part on your microscope, and practice using each part as it is described. In addition, be sure to fill in the parts on the microscope diagram pictured in Figure 3.1.

PARTS OF A COMPOUND LIGHT MICROSCOPE

1. **Occular(s):** The occular of a microscope is the part that you look through to view a specimen. In some cases, microscopes are *monocular,* meaning they only have one occular. Our microscopes are *binocular* because they have two. Notice that you can adjust the distance between occulars, which you will need to do later when viewing a specimen with both eyes open.

2. **Nosepiece:** The nosepiece is a revolving structure that holds the four objective lenses. Practice rotating the nosepiece, and notice how an objective will "click" into place when it is ready to be used.

3. **Arm:** The arm is a major support structure for the microscope, and is also where you should always have one hand when carrying your microscope (the other should be holding the **base**). Notice that the arm has a handle for easy carrying.

4. **Objective Lenses:** The objective lenses provide different levels of magnification when viewing specimens with the microscope. There are four objectives on our microscopes:

   a. **Scanning Objective:** This is the smallest objective, and is used to scan your specimen before focusing at higher powers. Notice how the scanning objective has a **4×** label. This refers to the magnifying power of the objective.

   b. **Low-Power Objective:** This objective has a **10×** power, which means it provides a greater magnification than scanning power.

   c. **High-Power Objective:** This objective has a **40×** power. Unlike the scanning and low-power objectives, the high-power objective is long enough that it can potentially contact your slide and/or stage if you are not careful. For this reason *only use the fine-focus knobs when focusing with the high-power objective!*

   d. **Oil Immersion Objective:** This is the most powerful objective on the microscope, and has a **100×** power. Because the lens is so powerful, it requires a drop of oil to be placed on a slide to help focus the light further before it can be viewed. Never switch to this objective without first placing a drop of oil on your slide!

5. **Stage:** The stage is the black square beneath the objective lenses that holds your slide.

FIGURE 3.1

Important Parts of a Compound Light Microscope

© Van-Griner, LLC

6. **Stage Clip:** This metal, claw-shaped structure can be moved using the small knob on top to make room for your slide. Practice fitting a slide into the stage clip by moving the clip outward.

7. **Mechanical Stage Adjustment:** There are two knobs below the stage that can be turned to move the slide forward/backward and side to side across the stage. Practice turning these knobs and notice how you can move the slide without having to take it out of the stage clip.

8. **Focus Knobs:** These knobs can be found on both sides of the microscope below the arm. Each knob is a pair of knobs, which can be turned independently:

   a. **Coarse-Adjustment Knob:** This is the knob that raises or lowers the stage to bring an object into approximate focus. You should *only* use this knob when viewing a specimen with the scanning or low-power objectives!

   b. **Fine-Adjustment Knob:** This knob is on the outside of the coarse-adjustment knob, and brings an object into final focus. Use this knob when viewing specimens with high-power and oil-immersion objectives.

9. **Condenser:** This structure gathers light from the light source and directs it toward the specimen being viewed.

10. **Diaphragm Control:** This dial can be found just below the stage at the front of the microscope. It controls the amount of light that passes through the condenser.

11. **Light Source:** This lamp shines light through the specimen being viewed.

12. **Illumination Control:** This dial is at the base of the microscope, and controls the intensity of the light source.

13. **Base:** The bottom portion of the microscope. Always have one hand on the base when carrying the microscope!

PROCEDURE 3.2: FOCUSING THE COMPOUND LIGHT MICROSCOPE

Your instructor may give you verbal instructions on how to focus the microscope properly. You may also follow the following guidelines when using the microscope to view a slide:

1. Make sure that you have your nosepiece turned so that the *scanning* objective is set in place.

2. Take a glass slide from the slide box and fit it into your stage clip, making sure the slide is right-side-up (you should be able to read the slide label).

3. Use the mechanical stage adjustment knobs to center your slide so that the middle of the slide is directly below the scanning objective.

4. Use the coarse-adjustment knobs to slowly raise the stage and bring your specimen into view as you look through the occulars.

5. Once you have your specimen in focus at the scanning power, practice using the *diaphragm control* and *illumination control* to adjust the amount and intensity of the light that illuminates your specimen. You can also practice adjusting the width of the occulars to fit the distance between your eyes.

6. When advancing to a higher power, you will only need to use the fine-focus knobs if you already have your specimen in focus at a lower power. This is because our compound light microscopes are **parfocal,** meaning they stay in focus when switching to higher-powered objectives. Practice moving to *low-power* (10×) and notice how your specimen stays in focus.

7. Once you have your specimen in focus with the *low-power* objective, carefully switch to the *high-power* objective. Remember to *only use fine-focus at this point!*

8. When you have finished making observations at *high-power,* turn the nosepiece so that the *scanning* objective is back in place. ***Do not*** switch to *oil immersion* unless your instructor provides further instructions on how to prepare an oil immersion slide.

PROCEDURE 3.3: DETERMINING THE TOTAL MAGNIFICATION OF A MICROSCOPE

A microscope is considered to be *compound* if it contains more than one lens that magnifies a specimen. When looking at the magnification value of each **objective lens,** students often mistakenly believe that this value represents the actual amount that their specimen is being magnified. For instance, if you had the microscope set at the *low-power* objective, the specimen being viewed would not be magnified by only 10× because there is a second magnifying lens in the **occular.** For our microscope, the occular lens is a 10× lens. Therefore, if you had the microscope set at the *low-power* objective, the *total magnification* of the microscope at that setting would be 10 × 10, or occular magnification × objective magnification. Complete the following table for the *total magnification* values that correspond to each of the four objective lens settings of your microscope.

Total Magnifications for Four Objectives

TABLE 3.1			
OBJECTIVE LENS	**OBJECTIVE MAGNIFICATION**	**OCCULAR MAGNIFICATION**	**TOTAL MAGNIFICATION**
Scanning			
Low-Power			
High-Power			
Oil Immersion			

PROCEDURE 3.4: MEASURING THE DEPTH OF FIELD OF A MICROSCOPE

The *depth of field* of a microscope is the thickness of the object being viewed when in focus. In general, when switching to more powerful objectives, the depth of field shrinks. Refer to the following instructions to determine how depth of field varies at different magnifications.

1. Obtain a slide that is labeled "colored threads." Set your microscope to the scanning objective and place the slide on your microscope stage within the stage clip.

2. Bring your slide into focus using the coarse adjustment knob. Use the mechanical stage adjustment knob to focus on the area where the three threads overlap.

3. Switch to low-power and try to determine the order of the threads from top to bottom.

4. Carefully switch to high-power and reexamine the threads to try again to determine the order of the threads from top to bottom.

At which objective (low-power or high-power) can more than one thread be in focus at once? What does this indicate about the size of the depth of field at low- and high-power objectives?

PROCEDURE 3.5: MAKING A WET MOUNT OF A LIVE SPECIMEN

In the labs to come, we will be studying different types of cells under the microscope. For live cell preparations, it is often necessary to make a *wet mount,* which is a type of slide containing water and a coverslip. Refer to the following instructions and Figure 3.2 to make and view a wet mount:

1. Obtain a clean glass slide and use a plastic pipette to place a drop of water on the slide from a beaker containing *Elodea* plants.

2. Use the forceps to remove a small piece of leaf from the *Elodea* plant and place it near the center of the drop of water on your slide.

3. Obtain a square coverslip and lower it onto the slide at an angle to avoid air bubbles. Make sure that the coverslip is lined up with the edges of the slide.

4. Observe the slide under your microscope at scanning power, and proceed to view the slide at low and high powers.

5. In the circles below, sketch your specimen as it appears under scanning and high-power objectives.

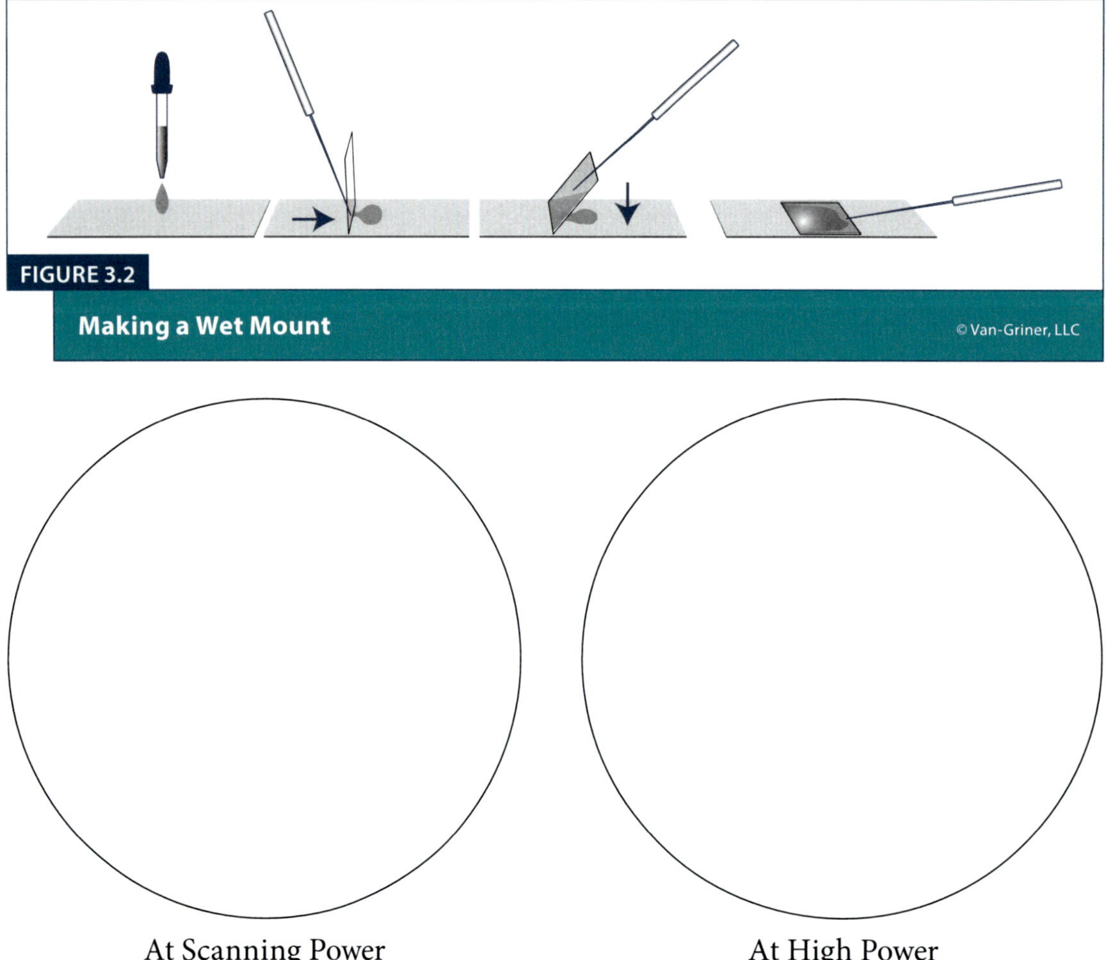

FIGURE 3.2

Making a Wet Mount © Van-Griner, LLC

At Scanning Power

At High Power

4

THE CHEMICAL BUILDING BLOCKS OF LIFE

OBJECTIVES

By the end of this lab exercise, students should be able to

- list and describe the four major types of macromolecules in cells;
- explain the reactions that cause polymers to form and break down;
- perform experiments to test for different macromolecules and their subunits; and
- analyze and explain the results of different experiments to verify the presence of specific macromolecules.

INTRODUCTION

When people talk about the prospect of life existing on other planets, they often specify that life must be "carbon-based." The element carbon forms the backbone of the building blocks that make up cells. Carbon can form chemical bonds with up to four atoms, including those of different elements like nitrogen, oxygen, and phosphorus. Carbon also has the ability to form ringed structures with other carbon atoms (Figure 4.1). Carbon is thus a versatile element, and this accounts for the variability in structure and function of the different building blocks that make up cells.

Methane

Leucine

Cholesterol

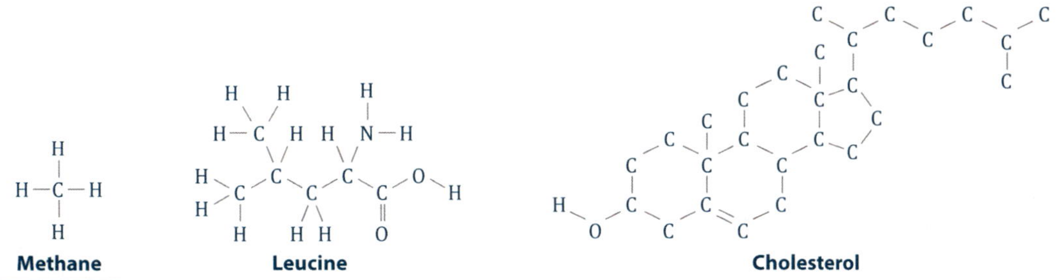

FIGURE 4.1

Common Molecules That Contain Carbon. Methane is the simplest carbon molecule. Leucine is an amino acid, which is a type of building block that forms proteins. Cholesterol is a steroid, which is a type of molecule composed of four carbon rings.

There are four major types of **macromolecules** that form the chemical composition of cells: proteins, carbohydrates, lipids, and nucleic acids. A macromolecule is a large molecule containing many atoms. Many biological macromolecules (biomolecules) exist as **polymers,** which are chains or smaller subunits. If a polymer were a string of beads, each bead would represent one subunit. In order for subunits to link together to form a polymer chain, a molecule of water must be removed. This is known as a **dehydration reaction** (see Figure 4.2).

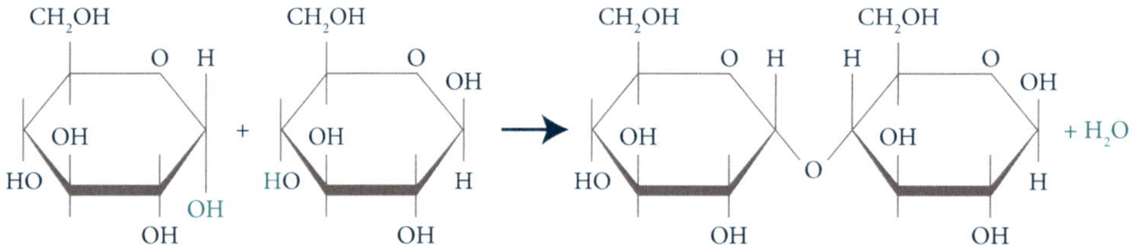

FIGURE 4.2

Dehydration Reaction. In a dehydration reaction, two hydrogen atoms and one oxygen (which make up one water molecule) are removed from two subunits to link them together.

As you might expect, organisms must often break apart the subunits of a polymer in order to obtain energy or to recycle chemical subunits. In a **hydrolysis** reaction, a molecule of water is added to a polymer to break it into subunits (see Figure 4.3). Hydrolysis and dehydration reactions are thus effectively opposite reactions. In this lab, we will be performing experimental tests to determine the presence of macromolecules and the subunits that make them up.

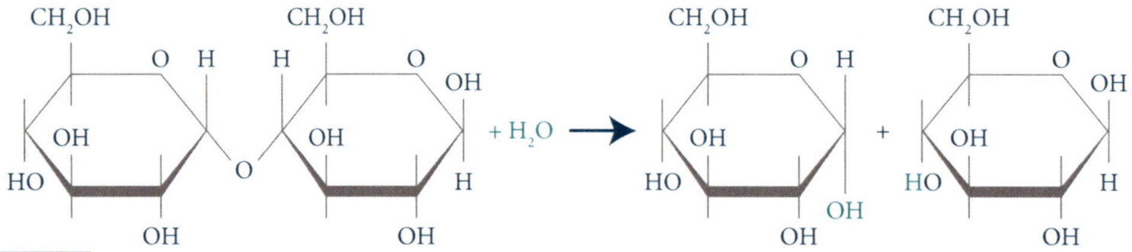

FIGURE 4.3

Hydrolysis Reaction. A hydrolysis reaction involves adding a molecule of water to break apart two subunits that are joined together.

PROCEDURE 4.1: TESTING FOR THE PRESENCE OF PROTEINS

Of the biomolecules found in living cells, none are more diverse than proteins. Proteins perform many functions in cells, including providing structural support, catalyzing reactions, transporting cargo around the cell and body, defending against pathogens, and many more. Part of the reason proteins are such a diverse type of molecule is due to their complex structure. Proteins are polymers (chains) of building blocks called **amino acids.** There are 20 different amino acids, which can chain together in any order. Two amino acids can be joined together by a dehydration reaction to make a **peptide,** which is a simple chain of two amino acids. The bond holding the amino acids together is called a **peptide bond.** When more than two amino acids bond together to make a chain, the resulting polymer is called a **polypeptide** (see Figure 4.4).

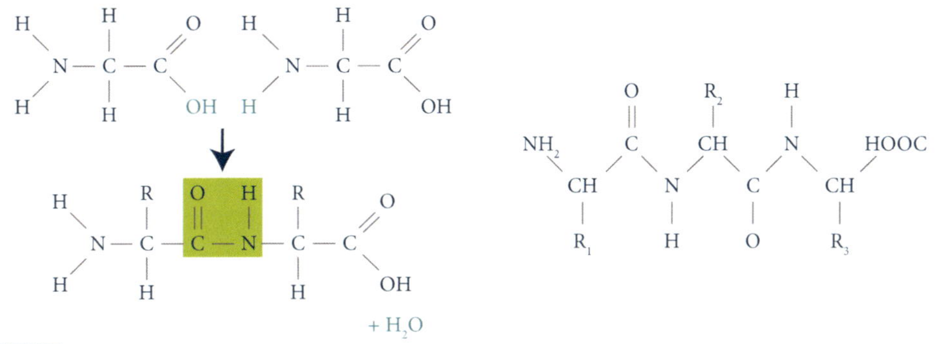

FIGURE 4.4 Peptide Bond   Polypeptide Bond

Amino Acids Forming Peptides and Polypeptides

Thinking Critically

Given the important functions proteins play in the body, it is no surprise that many people (especially athletes) enhance their daily protein intake by taking protein supplements. Whey protein is a popular source of protein used in supplements because it absorbs relatively quickly in the body. Some whey protein supplements are called *hydrolysates,* because they contain proteins that have been broken down by *hydrolysis* reactions. What advantage do you think *hydrolysate* protein supplements might provide over regular proteins supplements? Discuss with your lab partners and write your response below.

In order to test for the presence of proteins, an indicator solution called *Biuret reagent,* which contains dilute copper sulfate ($CuSO_4$) is used. When combined with peptides or polypeptides, the reagent turns a pink or purple color, depending on the concentration of peptide bonds (see Figure 4.5).

To begin, gather four clean test tubes and label them 1–4. For each test tube, use your small ruler to mark 1 cm up from the bottom of the tube and use your wax pencil to mark a straight line indicating the 1-cm mark (See Figure 4.6). This line represents the fill line for the volume of solution to be added to each test tube. Alternatively, your instructor may have you use transfer pipettes to add solutions to the tubes.

To each tube, add solutions from Table 4.1 to the 1-cm mark. Once each tube has been filled with its specific solution, add **5 drops of Biuret reagent** to each tube and gently swirl the tube to mix. After adding the reagent to each tube, observe any color change and complete the data columns in Table 4.1.

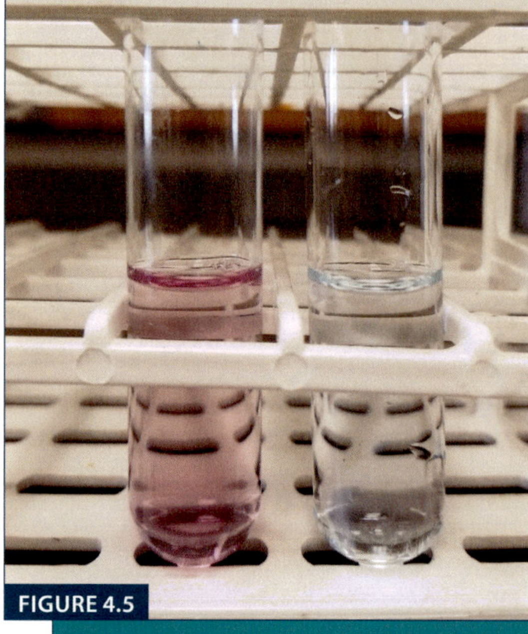

FIGURE 4.5

Test Tubes Containing Solutions with and without Peptide Bonds (Proteins)

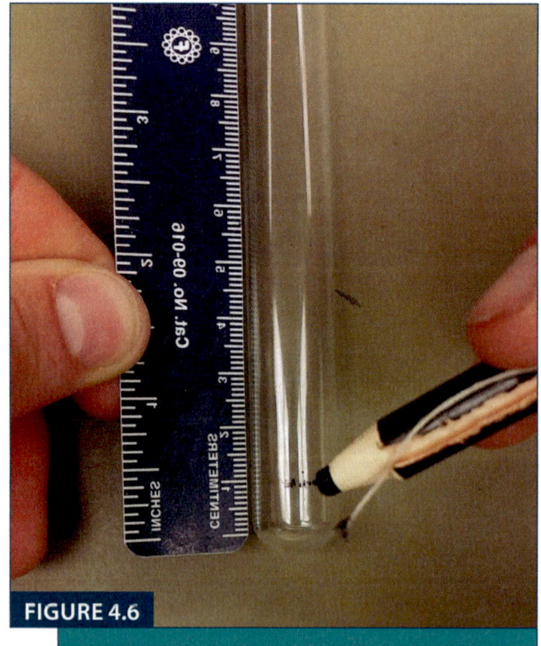

FIGURE 4.6

Marking the Fill Lines for the Experimental Test Tubes

Biuret Test for the Presence of Protein

TABLE 4.1			
TEST TUBE	**CONTENTS**	**FINAL COLOR**	**PROTEIN PRESENCE (+ OR −)**
1	Distilled Water		
2	Albumin		
3	Pepsin		
4	Glucose		

Based on your conclusions, what type of biomolecule are pepsin and albumin?

PROCEDURE 4.2: TESTING FOR THE PRESENCE OF CARBOHYDRATES

Carbohydrates are a type of biomolecule that are often immediately associated with energy, as glucose, a type of simple sugar, is the chief source of energy for most cells. In addition to fueling cells with energy, carbohydrates also play other important roles for different cells, including support and cell recognition. Like proteins, carbohydrates can exist as polymers of individual subunits. The subunits of carbohydrate chains are called **monosaccharides.** Glucose, for example, is a monosaccharide that most cells can break down to generate energy. Depending on the type of cell, carbohydrate monomers form polymers through dehydration reactions (see Figure 4.7). Two simple sugars can chain together to make a **disaccharide,** like table sugar (sucrose). When more than two simple sugars join together, the resulting structure is called a **polysaccharide.** In this lab, we will test for both simple sugars (glucose) and polysaccharides (starch).

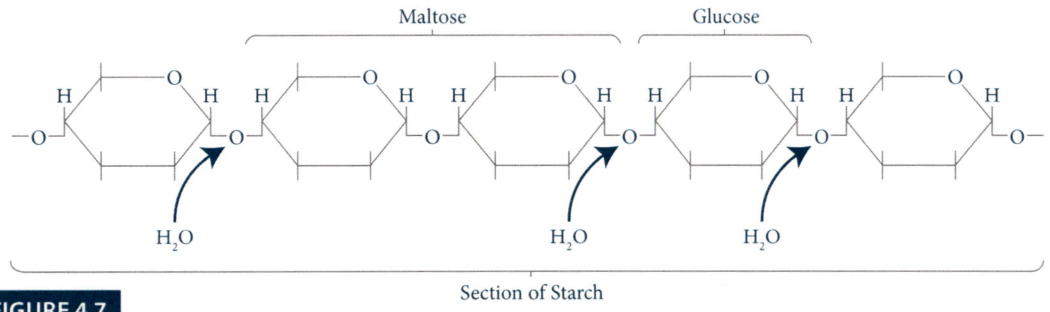

FIGURE 4.7

Glucose. Glucose can function in cells as an individual subunit or as chain of two or more subunits.

TESTING FOR GLUCOSE AND MALTOSE

As you can see from Figure 4.7, two glucose monomers can link together to form maltose, which is a disaccharide. *Benedict's reagent* is useful when trying to detect glucose, as copper ions in the reagent will react with glucose molecules to cause a color change, which can occur more rapidly at higher temperatures. The color of the solution will vary depending on how much glucose is present. Use the following table to assist in your conclusions at the end of the experiment.

How Benedict's Reagent Reacts with Glucose

TABLE 4.2	
SUBSTANCE TESTED	**COLOR CHANGE**
Distilled Water	None
Dilute (Low) Glucose Solution	Green
Moderate Glucose Solution	Orange
Concentrated (High) Glucose Solution	Red

To begin the experiment, gather five clean test tubes and label them 1–5. As you did in Procedure 4.1, measure 1 cm from the bottom of each test tube with a wax pencil (unless you are using transfer pipettes to measure out volumes of solutions). Add each of the appropriate experimental substances listed in Table 4.3 to the 1-cm mark of each tube. Next, add **5 drops of Benedict's reagent** to each test tube, then carry your test tube rack to one of the hot plates located on the side bench. Use the metal test tube clamps from your bin to transfer each tube into the beaker containing the boiling water on the hot plate. Be sure to *slowly* transfer each tube!

Allow the tubes to sit in the boiling water for **30 seconds.** Slowly and carefully use your test tube clamp to transfer the tubes back into your test tube rack, then take the rack back to your table to analyze your results. Fill in the data columns in Table 4.3.

Benedict's Test for the Presence of Glucose

TABLE 4.3			
TEST TUBE	**CONTENTS**	**COLOR CHANGE**	**GLUCOSE PRESENCE (NONE, LOW, MEDIUM, OR HIGH)**
1	Distilled Water		
2	Glucose Solution		
3	Soda		
4	Juice		
5	Starch		

1. Did your soda sample contain glucose? _______________________________

2. If the first ingredient listed on a soda label were *sugar* (sucrose), why would this test positive for Benedict's reagent?

____

____

____

TESTING FOR STARCH

When people think of "starchy" foods, they often think of potatoes. Many people attempt to limit the amount of starch in their diet because of an association between starch and calories. As you can see from Figure 4.7, starch is composed of multiple glucose monomers linked together in a chain. In order for cells to utilize the glucose monomers for energy, the bonds forming the starch polymer must be broken.

Iodine is often used to test for the presence of starch. Iodine solution, which has an amber color, reacts with starch molecules to produce a **dark (black with a slight blue tint) color.** We can use this color change to test for starch in different sample solutions.

To begin, gather five clean test tubes and label them 1–5 as in the previous experiment. Measure 1 cm from the bottom of each tube with your wax pencil, and fill each test tube up to the 1 cm mark with the appropriate substances listed in Table 4.4. Once each tube is filled with its appropriate substance, add **5 drops of iodine solution** to each tube and note the color change that takes place. Fill in the data columns in Table 4.4 after you have carried out the experiment.

Iodine Test for Starch

TABLE 4.4			
TEST TUBE	**CONTENTS**	**FINAL COLOR OF SOLUTION**	**STARCH PRESENT (+ OR − BASED ON COLOR CHANGE)**
1	Distilled Water		
2	Starch Suspension		
3	Onion Juice		
4	Potato Juice		
5	Glucose Solution		

1. Based on your results from Table 4.4, what can you conclude about the starch content of potato and onion cells?

2. Sometimes experiments like the iodine test for starch can have limitations. If you wanted to test for the starch content of *cola,* explain why this test might not provide accurate results.

TESTING FOR STARCH UNDER A MICROSCOPE: COMPARING STARCH STORAGE IN ONIONS AND POTATOES

In Lab 3 you will get a chance to observe different types of cells under the microscope. As you will discover, some cells have specialized, membrane-bound compartments called *organelles* that are associated with specific functions. Given that you just discovered that potato juice contains starch, we will now have a chance to observe how potato cells contain starch molecules. **Amyloplasts** are compartments in plant cells that house starch molecules (see Figure 4.8). While amyloplasts normally do not contain any colored pigments, staining them with iodine makes them highly visible under a microscope.

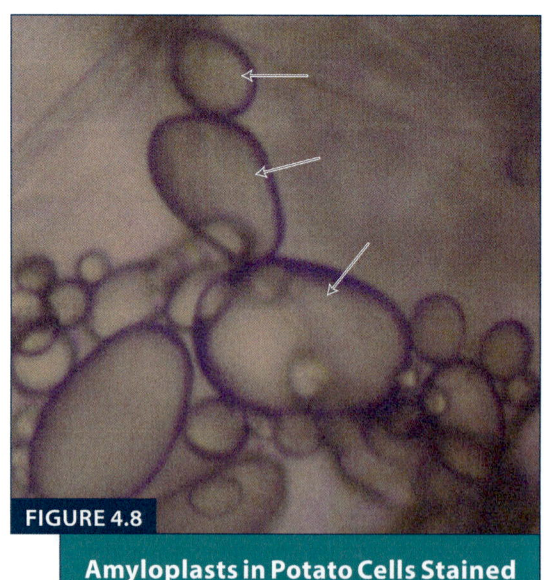

FIGURE 4.8

Amyloplasts in Potato Cells Stained with Iodine

To begin this procedure, you will need a compound light microscope. Obtain a clean, glass slide and place 1 drop of water from a beaker on the side bench on the center of the glass slide. Using a scalpel, **carefully** shave off a piece of potato that is *about as thin as paper* and no wider or longer than the drop of water on the glass slide. Next, gently place the piece of potato in the drop of water and carefully add a glass coverslip onto the drop of water. You can now view your specimen under the microscope. Follow the microscope procedure you learned from Lab 2. Once at 40× magnification, observe the boundaries of each potato cell, and notice that the cells form a mosaic pattern

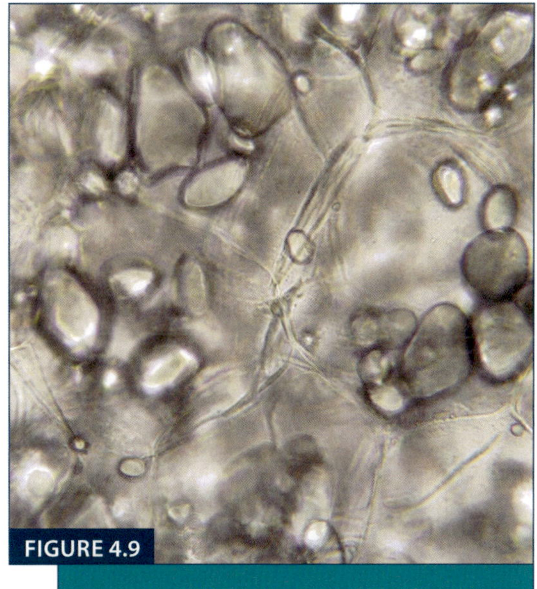

FIGURE 4.9

Potato Cells at 400× Magnification

(see Figure 4.9). Without a stain, it is difficult to identify the structures in the cell. Before staining, we will prepare a similar wet-mount sample with a piece of onion.

Obtain a second clean, glass slide and add a drop of water in the center. Peel (using your fingernails or the scalpel) a piece of onion from a single, transparent layer of onion skin and place it on your slide within the drop of water. As with the potato, the piece of onion skin should be paper-thin and should not be wider or longer than the drop of water on the slide. Once you have added the cover slip onto the drop of water containing the onion skin, you may begin viewing the onion under the microscope. Notice that onion cells also form a mosaic pattern of cells that bear a similar size and shape (see Figure 4.10).

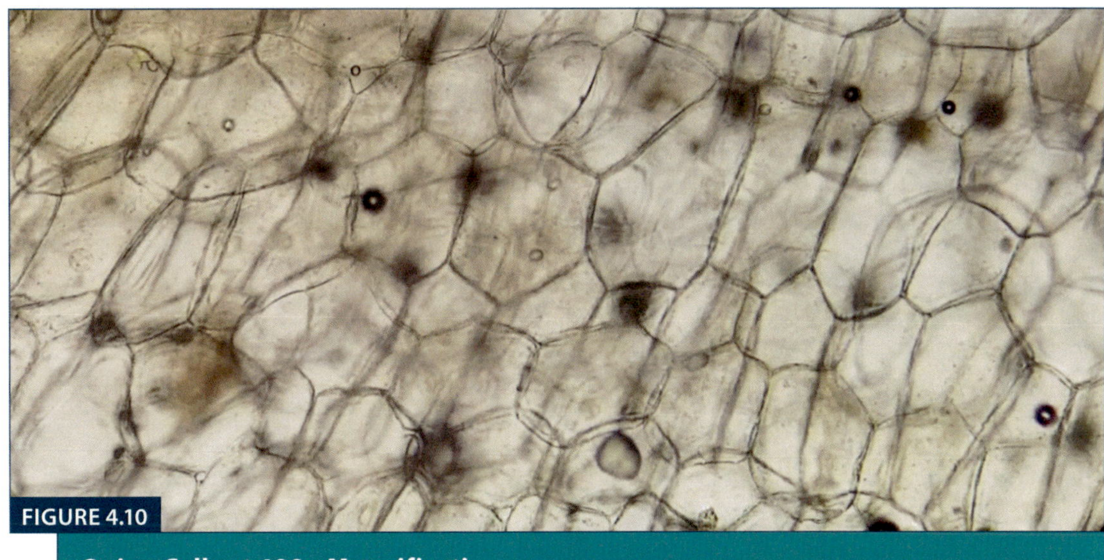

FIGURE 4.10

Onion Cells at 400× Magnification

After viewing the onion and potato cells, we will now stain the cells with iodine to see which contains more starch. To do this, we will employ a technique known as *wicking.* The goal is to have iodine contact the cells under the coverslip, but we do not want to remove the cover slip or the slide from within the stage clip. Instead, use the mechanical stage adjustment knobs to bring the slide away from the objective (and closer to you). Once the slide is about two inches from the objective, add a single drop of iodine to *one side* of the cover slip (see Figure 4.11). Then use a small piece of paper towel to soak up some of the water from under the coverslip on the opposite side from where you added the iodine (see Figure 4.12). Due to the properties of water, the iodine will be pulled under the coverslip, and will eventually contact the cells. Place the slide back under the objective and begin to view the cells as you did before. *Repeat this procedure for both the onion and potato.*

FIGURE 4.11

Adding a Drop of Iodine to a Slide Sample

FIGURE 4.12

Using a Paper Towel Piece to Wick the Fluid under the Coverslip

Once you have observed the cells after they have been stained, write your conclusions about which type of cell (onion or potato) contains more starch based on your observations.

CONCLUSIONS: COMPARING STARCH STORAGE IN ONIONS AND POTATOES

DESIGNING AN EXPERIMENT TO TEST THE EFFECTS OF AMYLASE

In Lab 5, you will learn the details of a type of protein called an *enzyme*. Enzymes are capable of catalyzing specific biological reactions, which makes the reactions occur much faster than they would outside of a cell. **Amylase** is a type of enzyme produced in animals that assists in the breakdown of starch molecules into monosaccharides. To conclude the procedure on carbohydrates, you will design your own experiment to test the activity of the amylase enzyme.

Refer to the carbohydrate procedures you completed to help you design an experiment that would verify that amylase can hydrolyze starch. Discuss with your lab partners as to how you would conduct the experiment, and then write a brief outline of how you will carry out the experiment in the space below.

EXPERIMENTAL OUTLINE: TESTING THE EFFECTS OF AMYLASE

Before you carry out your experiment, check with your instructor to make sure that your experiment is sound. Once you are given the OK, carry out your experiment and write your conclusions in the space below.

CONCLUSIONS: TESTING THE EFFECTS OF AMYLASE

PROCEDURE 4.3: TESTING FOR THE PRESENCE OF LIPIDS —

Lipids are a large and diverse group of macromolecules, which contain fats, oils, steroids, and waxes. While lipids vary in function from energy storage to forming cell membranes, they all share the characteristic of being **hydrophobic,** which literally means "water fearing." Hydrophobic molecules are formed from non-polar covalent bonds, which means that the atoms in the molecules do not form hydrogen bonds with water. Figure 4.13 displays the chemical structure of structure of two common lipids.

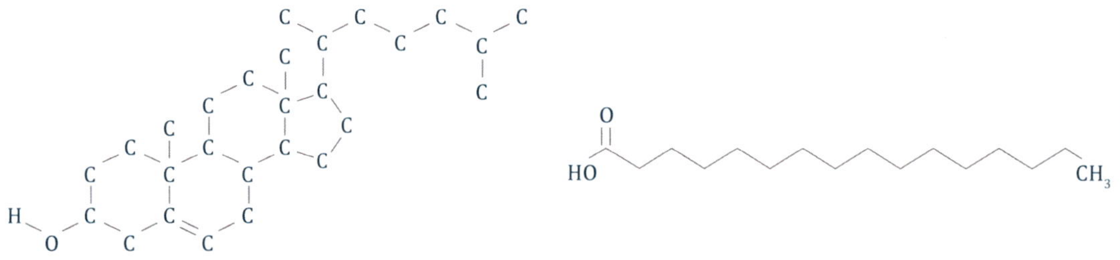

Cholesterol

Palmitic Acid

FIGURE 4.13

Chemical Structures of Cholesterol (a Steroid) and Palmitic Acid (a Fatty Acid)

TESTING LIPIDS USING SUDAN RED

In chemistry, **solutes** are compounds that can dissolve in **solvents.** Water is often called "the universal solvent," because water molecules can easily form hydrogen bonds with polar molecules to dissolve them. As you may know, lipids like oil do not dissolve in water. Lipids do, however, dissolve in other lipids. Chemists often use the phrase "like dissolves like" to help remember that compounds, like lipids, dissolve in other compounds of similar chemical makeup.

In this experiment, you will be working with a type of dye called *sudan red*. Sudan red is a lipid-based dye, which will only dissolve in certain solutions.

To begin, gather and label two test tubes, and use the following instructions to fill the tubes:

> **Test Tube 1:** Fill with **5 mL of distilled water**

> **Test Tube 2:** Fill with **5 mL of vegetable oil**

Next, add **5 drops of sudan red** to each tube and observe the differences between the two tubes.

1. Did the sudan red dye mix with the water in Test Tube 1? _______________

2. Did the sudan red dye mix with the oil in Test Tube 2? _______________

3. If you did not have vegetable oil available, what other substance might you be able to use to dissolve sudan red? _______________

EMULSIFYING LIPIDS

When you consume lipids like fats and oils, your body must take extra steps in the digestive process so that the hydrophobic molecules can be properly absorbed into the tissues. Because lipids are insoluble (do not dissolve) in water, they must be exposed to a substance that is capable of breaking them down so that they can be absorbed effectively. Lipids can be broken down into smaller parts in the process of **emulsification.** In the human body, bile is a substance that is produced in order to emulsify lipids. In this procedure, we will test the effects of adding an emulsifier to a lipid solution.

Obtain two clean test tubes. Use a transfer pipette to fill each tube with 5 mL of distilled water and 5 mL of vegetable oil. Number the test tubes 1 and 2.

To Test Tube 1, add 1 mL of **bile salts** and allow the solutions to react for a minute or two.

Using a plastic pipette, transfer a few drops from each test tube near the boundary between the oil and water layers (see Figure 4.14) and place a drop on two separate microscope slides. Add coverslips to the slides and prepare to view them at 4× and 10× magnifications.

Briefly describe your observations in both slide samples, including how you are able to determine that the bile salts emulsified the lipid droplets.

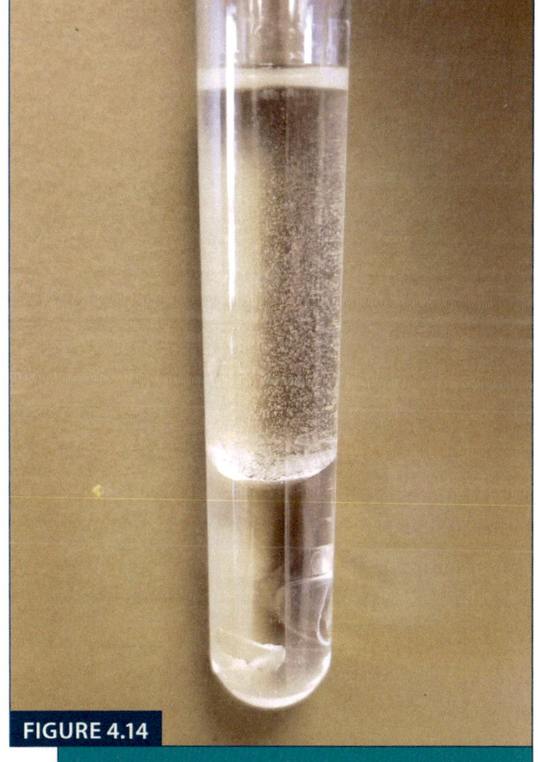

FIGURE 4.14

A Test Tube Containing Oil, Water, and Bile Salts. The bile salts are emulsifying the oil (top layer).

5

MEASURING ENZYME ACTIVITY

OBJECTIVES

By the end of this lab exercise, students should be able to

- explain how the structure of enzymes relates to their function;
- differentiate between enzymes, substrates, and products;
- describe the reaction that is catalyzed by the enzyme *catalase;* and
- explain the factors that are capable of influencing enzyme activity by conducting experiments measuring *catalase* activity.

INTRODUCTION

In order for cells to function in living systems, they require a host of reactions that help them build and break down molecules. Most of these reactions occur too slowly on their own to keep up with the metabolic demands of cells. It is therefore essential for cells to contain **enzymes** to help make reactions occur. Enzymes are proteins that catalyze reactions by lowering the *activation energy* of the reaction, which is the amount of energy input required for a reaction (see Figure 5.1). This allows a specific reaction to occur more readily in a cell.

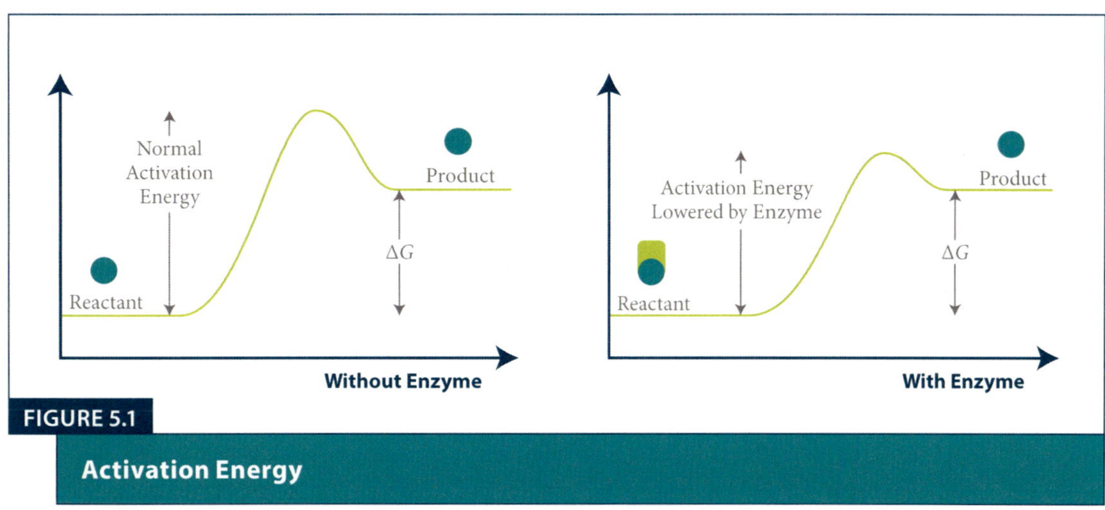

FIGURE 5.1

Activation Energy

Enzymes, like all proteins, take on a three-dimensional shape based on the amino acid building blocks that make them up. In order for a given enzyme to catalyze a specific reaction, it must bind to a **substrate,** which is a molecule that binds to the enzyme during an enzymatic reaction. Substrates are able to bind to enzymes at **active sites** similar to how a specific key is able to fit into a specific lock. Sometimes enzymes may bind to one substrate and break the substrate into two new **product** molecules, which are the molecules that result from an enzymatic reaction. Other enzymes may bind to multiple substrates at once to produce a single product from two smaller molecules. Enzymes are thus capable of breaking, forming, or rearranging the chemical bonds that make up specific compounds (see Figure 5.2).

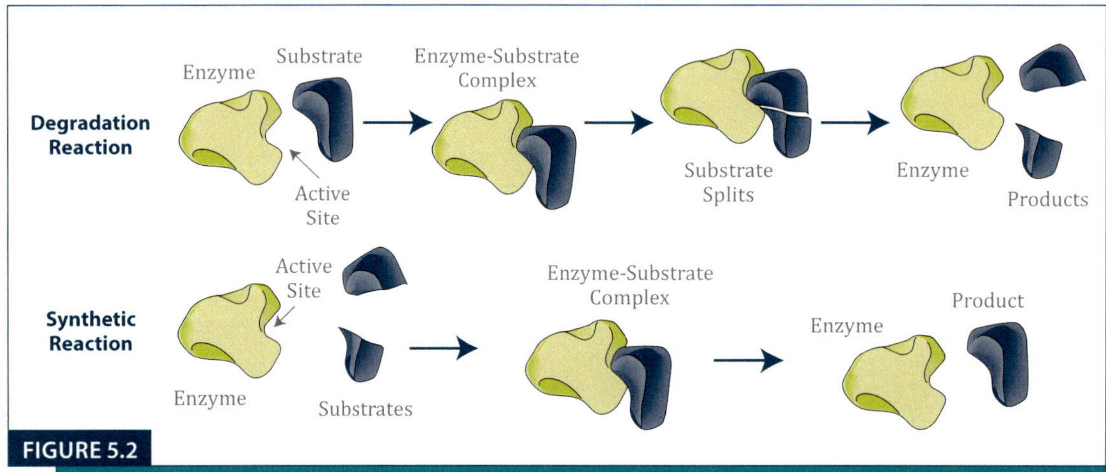

FIGURE 5.2

Enzyme Reaction. Enzymes can bind to substrates to build (synthesize) larger molecules, or break down (degrade) large molecules into smaller ones.

Much like locks are typically very selective in terms of which keys they fit, enzymes are often very selective in terms of the substrate to which they can bind. As you learned in Lab 4, proteins have specific chemical properties, and are thus subject to changes in their environment. As you hypothesize about how our experimental enzyme will be affected under different conditions, remember that an active site, just like a lock, can only function if it has the correct shape.

In this lab, we will be measuring the activity of an enzyme called catalase. Catalase is an enzyme produced in most cells, and it functions by binding to hydrogen peroxide (H_2O_2) and breaking it down into water (H_2O) and oxygen (O_2). The overall reaction can be summarized as

$$\text{Catalase} + 2H_2O_2 \longrightarrow 2H_2O + O_2.$$

Enzyme · · · · · Substrate · · · · · · · · · · · Products

Thinking Critically

When have you ever used hydrogen peroxide (H_2O_2) in the past? Given what you know about hydrogen peroxide, explain why you think it is important for cells to have a way of breaking this compound down.

PROCEDURE 5.1: ASSESSING THE CATALYTIC ACTIVITY OF CATALASE

In this lab, we will be combining the enzyme with its substrate in individual test tubes. This reaction should produce oxygen, which we will measure at specific intervals after the enzyme and substrate are combined (see Figure 5.3). After the appropriate solutions have been combined in a given test tube, allow the reaction to commence for **30 seconds,** and then take a measurement (in cm) of the bubble column that forms in the tube. It is important that you make sure to take your activity measurements in each test tube precisely 30 seconds after combining the reagents.

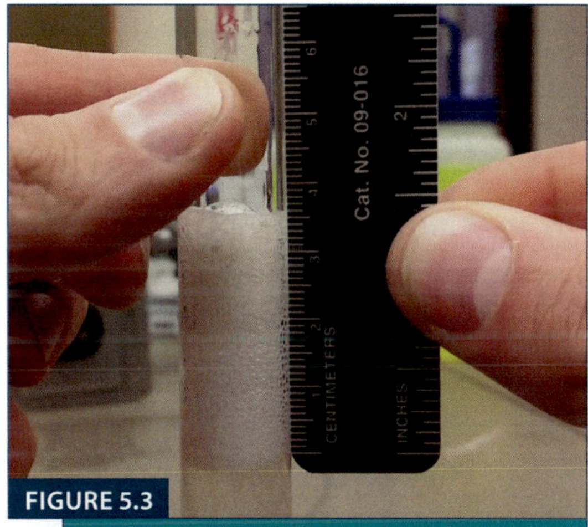

FIGURE 5.3

Measuring the Bubble Column. When peroxidase reacts with hydrogen peroxide in a test tube, a column of bubbles forms due to the production of oxygen gas, which can be measured with a ruler.

Thinking Critically

Discuss with your group members why it is important to keep the elapsed time for each test tube constant. Provide a brief explanation below of why it would be a mistake to combine the reagents of each tube, and then time them all together.

Before we begin to explore the ways in which enzyme activity can be influenced, we must first demonstrate that catalase will catalyze a reaction that we can measure. You will need to verify that combining catalase with hydrogen peroxide will produce a measurable amount of oxygen in a test tube. Before beginning, review the reagents that will be added to each of the four tubes in this experiment (see Table 5.1). Work with your lab partners to design a hypothesis as to how you think the reaction will proceed differently (if at all) in each test tube.

HYPOTHESIS: ASSESSING THE CATALYTIC ACTIVITY OF CATALASE

To begin your first experiment, obtain four test tubes and place them in your test tube rack. Label each tube from 1–4 using the wax pencil in your bin. Using the pipetting procedure that you learned from Lab 2, transfer the following reagents into each tube, and then use your plastic ruler to measure the height of the bubble column that forms after 30 seconds. Remember that you should only prepare and measure one tube at a time! Record your data in Table 5.2.

TABLE 5.1

TEST TUBE	TEST TUBE CONTENTS
1	1 mL of Catalase Solution 4 mL of Hydrogen Peroxide
2	1 mL of Distilled Water 4 mL of Hydrogen Peroxide
3	1 mL of Catalase Solution 4 mL of Distilled Water
4	1 mL of Lipase Solution 4 mL of Hydrogen Peroxide

TABLE 5.2

TEST TUBE	BUBBLE COLUMN HEIGHT (cm)
1	
2	
3	
4	

RESULTS: ASSESSING THE CATALYTIC ACTIVITY OF CATALASE

Briefly describe what you can conclude about the activity of the enzyme catalase, and state whether or not your hypothesis was supported.

What was/were the **independent variable(s)** in this experiment? What was/were the **dependent variables?**

PROCEDURE 5.2: DETERMINING HOW TEMPERATURE INFLUENCES ENZYMES

Recall that enzymes require specific, three-dimensional active sites that will bind to the substrate in the reaction that they are catalyzing. In this experiment, you will be examining the effect of temperature on catalase activity by storing samples of catalase at different temperatures. You will need five clean test tubes, which you can number 1–5 with your wax pencil.

You will add 1 mL of **catalase solution** to each tube, and you will then place the tubes at specific temperature stations located around the room. For each station, record the temperature using a thermometer. Use Table 5.3 to help you set up set up the first part of the experiment.

TABLE 5.3

TEST TUBE	TEST TUBE CONTENTS	TEMPERATURE LOCATION	
1	1 mL Catalase Solution	Ice Bath:	_____ °C
2	1 mL Catalase Solution	Refrigerator:	_____ °C
3	1 mL Catalase Solution	Desk (room temp.):	_____ °C
4	1 mL Catalase Solution	Warm Water Bath:	_____ °C
5	1 mL Catalase Solution	Boiling Water:	_____ °C

After you have added the catalase solution to each tube and placed it in the appropriate setting, allow your tubes to remain at each temperature station for **15 minutes.** As you wait, work with your lab partners to develop a hypothesis as to how you think the catalase reaction will differ (if at all) among the five test tubes.

HYPOTHESIS: DETERMINING HOW TEMPERATURE AFFECTS ENZYMES

Once each tube has been sitting at its specific temperature station for at least 15 minutes, you may begin to test the activity in each tube. It is important that you measure the activity in only one tube at a time! Choose one of the tubes and add **4 mL of hydrogen peroxide** to the tube. Wait **30 seconds** and measure the bubble column height in cm. Proceed with the remaining tubes until you have measured activity in all the tubes. Record your data in Table 5.4

TABLE 5.4

TEST TUBE	TEMPERATURE LOCATION	BUBBLE COLUMN HEIGHT (cm)
1	Ice Bath: _____ °C	
2	Refrigerator: _____ °C	
3	Desk (room temp.): _____ °C	
4	Warm Water Bath: _____ °C	
5	Boiling Water: _____ °C	

RESULTS: DETERMINING HOW TEMPERATURE AFFECTS ENZYMES

Briefly describe what you can conclude about how temperature affects catalase, and state whether or not your hypothesis was supported.

When an enzyme shape is permanently changed, it is said to be **denatured** (see Figure 5.4). In which of the tubes do you think your catalase enzyme was denatured? Explain why below.

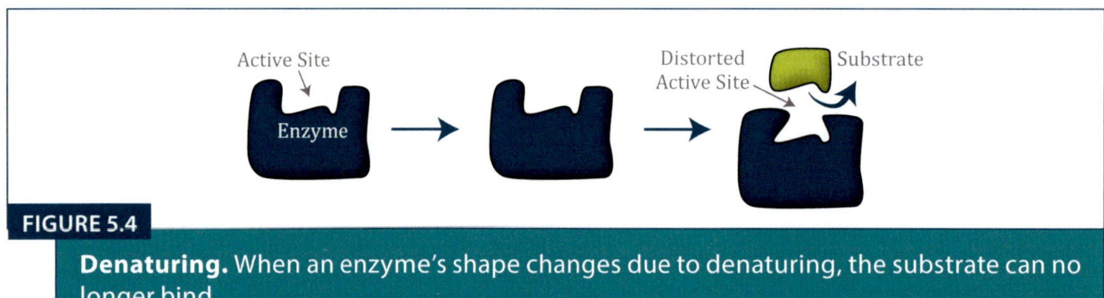

FIGURE 5.4

Denaturing. When an enzyme's shape changes due to denaturing, the substrate can no longer bind.

PROCEDURE 5.3: DETERMINING THE OPTIMUM pH ENVIRONMENT FOR CATALASE

In chemistry, the pH of a solution is a measure of its acidity or basicity. The pH scale ranges from 0-14, and measures the concentration of hydrogen ions present in a solution. The pH of pure water is 7.0, which is considered neutral. Any substance with a pH lower than 7.0 is considered an **acid,** and anything above 7.0 is considered a **base.**

Biomolecules, like all molecules, are held together by chemical bonds. Some of these bonds, called ionic bonds, can be disrupted by extremely acidic or basic conditions. Enzymes, which require specific three-dimensional shapes to bind to substrates, can have their active sites disrupted by extreme pH environments, leaving the enzyme denatured. Depending on the type of cell in an organism, an enzyme might be required to function at low, high, or neutral pH values. Most enzymes thus have an *optimum pH,* which is the pH level at which the enzyme functions best. Catalase, which is found in most organisms, tends to function best at pH values close to neutral.

Thinking Critically

Work with your lab partners to try to think of an enzyme in the human body that must function in an extremely acidic environment. Knowing what you now know about enzymes, what do you think would happen if other enzymes in your body were exposed to this type of environment? Can you think of a condition in the human body when an extreme pH environment might affect cells and enzymes that are not equipped to deal with it? Write your responses below.

To begin this procedure, collect five clean test tubes and place them in your test tube rack. Label each 1–5 with your wax pencil, and find the labeled pH solutions in your bin. You will need a separate pipette tip for each pH solution used! In each tube, you will first add **1 mL of catalase solution** and **2 mL** of a specific **pH solution.** Use Table 5.5 to help you fill each test tube **before** adding the substrate.

TABLE 5.5

TEST TUBE	TEST TUBE CONTENTS
1	1 mL Catalase Solution 2 mL pH 3 Solution
2	1 mL Catalase Solution 2 mL pH 5 Solution
3	1 mL Catalase Solution 2 mL pH 7 Solution
4	1 mL Catalase Solution 2 mL pH 9 Solution
5	1 mL Catalase Solution 2 mL pH 11 Solution

Once you have added the appropriate contents to each tube, let each tube sit for at least **3 minutes.** While you wait, discuss with your lab partners to determine a hypothesis as to how you think the activity will differ (if at all) among the five tubes.

HYPOTHESIS: DETERMINING THE OPTIMUM pH ENVIRONMENT OF CATALASE

After allowing your test tubes to sit for 3 minutes, you may begin to add **4 mL of hydrogen peroxide** to each tube and measure the bubble column after **30 seconds** as you did before. Remember to measure activity one tube at a time. Record your results in Table 5.6.

TABLE 5.6

TEST TUBE	pH ENVIRONMENT	BUBBLE COLUMN HEIGHT (cm)
1	3.0	
2	5.0	
3	7.0	
4	9.0	
5	11.0	

RESULTS: DETERMINING THE OPTIMUM pH ENVIRONMENT OF CATALASE

Briefly describe what you can conclude about the optimum pH environment for catalase, and state whether or not your hypothesis was supported.

We did not include pH environments below 3 or above 11 in this experiment, even though the scale ranges from 0–14. Do you think your conclusion would have changed if you included a test tube containing a pH value of 1.0 and one containing a pH value of 13.0? Briefly explain your reasoning below.

PROCEDURE 5.4: ESTABLISHING A RELATIONSHIP BETWEEN ENZYME CONCENTRATION AND CATALYTIC ACTIVITY

Imagine you have a flashlight that you wish to use in your house, but the flashlight requires a battery before it can function. You may want to put the light to use, but you will not be able to do so until you come across a battery, which may take time depending on how long it takes for you to come across the right battery for the job. It stands to reason that your chances of finding a battery would increase if you had many batteries scattered about your house, as you would be more likely, even by chance alone, to stumble across one. Because an enzyme must physically come into contact with its substrate in order for its reaction to be catalyzed, you might be able to speculate as to how increasing the amount of substrate, just like increasing the amount of batteries in your house, might affect how quickly a reaction will proceed. In this experiment, we will be keeping the substrate concentration constant in each trial, but will manipulate the concentration of catalase in each test tube.

Before you begin preparing your test tubes, work with your lab partners to develop a hypothesis as to how you think changing the concentration of catalase will affect its catalytic activity.

HYPOTHESIS: ESTABLISHING A RELATIONSHIP BETWEEN ENZYME CONCENTRATION AND CATALYTIC ACTIVITY

Begin this procedure by collecting four clean test tubes and placing them in your test tube rack. Label each tube 1–4 with your wax pencil. Fill each tube with catalase solution and hydrogen peroxide based on the volumes listed in Data Table 5.4. Combine the contents of each tube **one at a time** and allow the reaction to proceed for 30 seconds before measuring the bubble column height with your ruler. Record the bubble column height data for each tube in Table 5.7.

TABLE 5.7

TEST TUBE	CONTENTS	BUBBLE COLUMN HEIGHT (cm)
1	0.5 mL Catalase Solution 4 mL Hydrogen Peroxide	
2	1 mL Catalase Solution 4 mL Hydrogen Peroxide	
3	2 mL Catalase Solution 4 mL Hydrogen Peroxide	
4	3 mL Catalase Solution 4 mL Hydrogen Peroxide	

RESULTS: ESTABLISHING A RELATIONSHIP BETWEEN ENZYME CONCENTRATION AND CATALYTIC ACTIVITY

Briefly describe the relationship that you observed between the concentration of catalase and its ability to catalyze the breakdown of hydrogen peroxide, and state whether or not your hypothesis was supported.

Do you think that increasing the concentration of *substrate* in each tube (while keeping the concentration of enzyme constant) would have produced similar results? Briefly explain why/why not below.

PROCEDURE 5.5: EVALUATING THE CATALASE CONTENT OF DIFFERENT FOODS

Like animals, different plants can live in vastly different types of environments, and can be adapted in vastly different ways. Some plants may produce hydrogen peroxide in higher quantities than others, and may require more catalase to break it down. The readily-available fruits and vegetables that we buy in grocery stores are products of living plants, and thus contain variable amounts of catalase. Your instructor has liquified different fruits and vegetables to produce different solutions that you will be using to measure catalase activity. In this procedure, you will be designing your own experiment to determine the catalase content of the fruit and vegetable samples provided. Use the space below to outline how you will carry out your experiment. You may list the steps of your experiment (including volumes of solutions used) or summarize your experimental outline in a table. After you have developed your experimental outline, be sure to notify your instructor before proceeding to carry out the tests.

EXPERIMENTAL OUTLINE: EVALUATING THE CATALASE CONTENT OF DIFFERENT FOODS

Before beginning your experiment, work with your group members to develop a hypothesis as to how you think the catalase activity will differ (if at all) among the samples that you will test. ***Remember:*** A hypothesis, in some instances, can be an educated guess. Discuss anything you can think of about the plants that produce the fruits and vegetables in your experiment (where they live, how big they get, etc.) to help you form a hypothesis.

HYPOTHESIS: EVALUATING THE CATALASE CONTENT OF DIFFERENT FOODS

Now proceed with your experiment, and be sure to keep a note of your activity measurements for each trial. When you are finished, summarize your results and any conclusions you can draw (including whether or not your hypothesis was supported) in the space below.

RESULTS: EVALUATING THE CATALASE CONTENT OF DIFFERENT FOODS

6

EXAMINING CELL STRUCTURE AND FUNCTION

OBJECTIVES

By the end of this lab exercise, students should be able to

- describe the basic components of the *Cell Theory;*
- compare and contrast *eukaryotic* and *prokaryotic* cells;
- identify and describe the important organelles in animal and plant cells;
- prepare and view living cells under a microscope; and
- sample for living cells in pond water.

INTRODUCTION

Cells are often described as the fundamental units of life, as they are the smallest component of living things that are considered to be alive. Multicellular organisms like humans may be composed of trillions of cells, while organisms like bacteria are made of only a single cell. Many of the processes and phenomena that are studied in biology depend upon the study of cells, but knowing about one type of cell is not very helpful, as cells come in a vast array of shapes and sizes (see Figure 6.1).

As you learned in Laboratory 3, studying cells requires specialized equipment like compound light microscopes. Much of what we know about the nature of life, including important topics like disease transmission, has come from studying cells. Before humans developed the technology to study cells, many topics in biology were poorly understood. The advent of microscopy thus allowed biologists to study cells directly, which led to the branch of biology called *cellular biology.* Early cellular biologists worked to develop the *Cell Theory,* which states the following:

1. All living organisms are composed of one or more cells.
2. The cell is the basic unit of life.
3. All cells come from pre-existing cells.

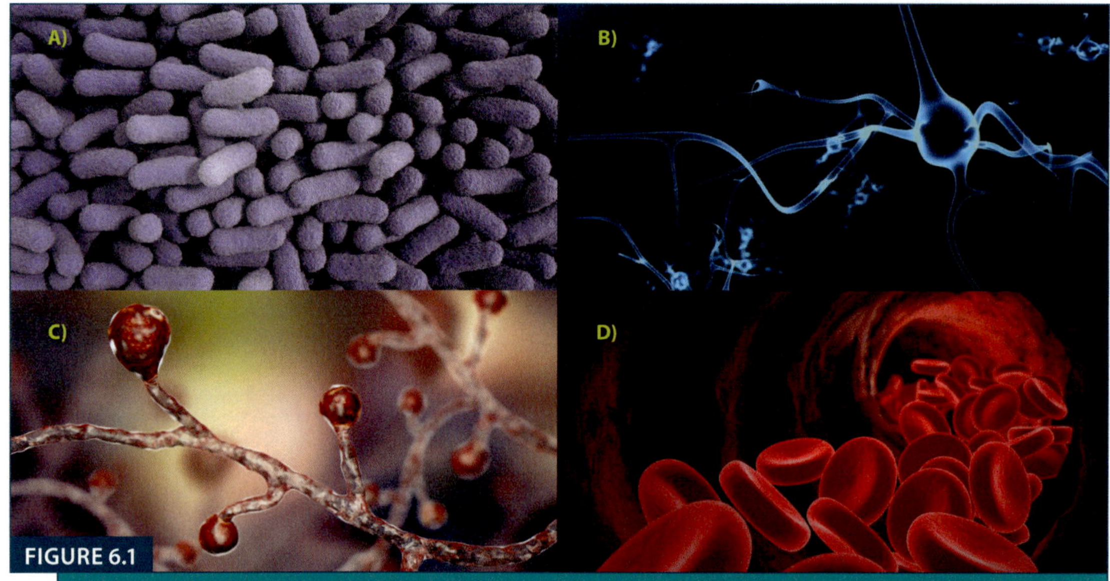

FIGURE 6.1

Cells Come in Many Shapes and Sizes. Shown here are renderings of A) bacterial cells, B) a nerve cell, C) fungal cells, and D) red blood cells.

While cells differ tremendously in form and function, there are certain characteristics that virtually all cells have in common. Other cellular characteristics and structures are unique to major groups of organisms like plants and animals. This lab will allow you to study live samples of different types of cells.

PROCEDURE 6.1: OBSERVING PROKARYOTIC CELLS

A **prokaryotic** cell lacks a nucleus and membrane-bound organelles. The two major groups of prokaryotic cells on earth are the *bacteria* and the *archaea*. Aside from lacking many of the complex cellular compartments that cells like your own have, prokaryotes are also, in general, many times smaller than most eukaryotic cells. Do not be mistaken by thinking that their relative simplicity makes them any "less evolved" or less successful than organisms like plants, animals, and fungi. Prokaryotes are extremely diverse, and can be found in just about every habitat on the planet.

In this lab, we will examine bacterial cells by preparing wet mount samples. Refer to Figure 6.2, which shows some of the notable cellular structures found a common bacterial cell like *E. coli*.

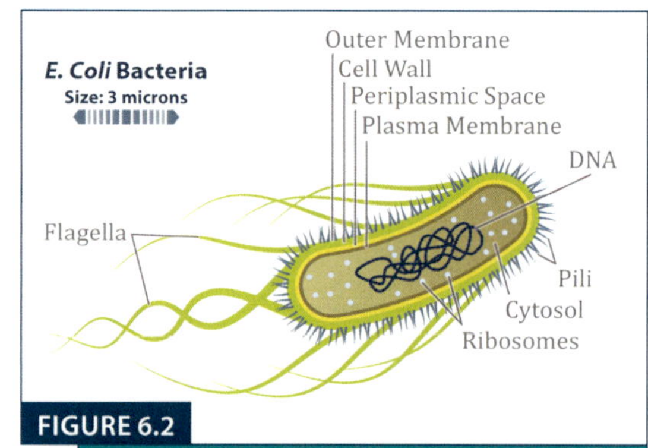

FIGURE 6.2

Common Structures Found in Bacterial Cells. Note that some structures are unique, but others (like ribosomes and a cell membrane) can be found in the eukaryotic cells featured later in the lab.

E. coli bacteria, as you might expect, are extremely small. To visualize a prokaryotic cell more easily, we will be preparing a slide of *cyanobacteria*. Cyanobacteria are unique among bacteria in that they, like plants and algae, can undergo a process called *photosynthesis* (studied in a later lab). This allows them to use solar energy to produce sugars and oxygen. Cyanobacteria are ideal for microscope viewing because they are much larger than other types of bacteria, and they also contain chlorophyll pigments, which give them a distinctive blue/green color (see Figure 6.3).

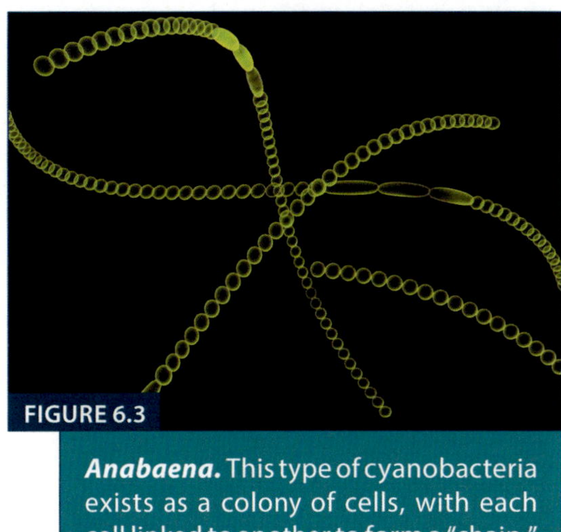

FIGURE 6.3

Anabaena. This type of cyanobacteria exists as a colony of cells, with each cell linked to another to form a "chain."

To make a wet mount slide preparation of cyanobacteria, obtain a clean glass slide and coverslip. Place a drop of water from one of the glass tubes containing cyanobacteria on the side bench (your instructor will tell you which type to prepare based on what is available in the lab) onto your glass slide and add the coverslip as you have done before. Begin observing your specimen under 4× power and advance to low and high powers.

Sketch your sample as you see it at 400× in the circle to the right. If possible, indicate in your drawing where an individual cell can be found within the colony of cells.

Cyanobacteria at 400× Magnification

PROCEDURE 6.2: COMPARING EUKARYOTIC CELLS: PLANT CELL AND ANIMAL CELLS

Eukaryotic cells contain nuclei (eukaryote means "true nucleus"). A nucleus, as seen in Figure 6.4 and Figure 6.5, is a centrally-located structure that contains a cell's DNA. Another key distinctions between eukaryotic and prokaryotic cells is the presence of membrane-bound **organelles.** Organelles are specialized compartments within eukaryotic cells that carry out specific tasks. Because organelles are able to function in specific ways, they are surrounded by membranes so that their reactions can be compartmentalized from the rest of the cell. Some of these compartments are responsible for generating energy, while others help break down unwanted substances. Plants and animals are both examples of eukaryotic organisms. Compare the cell structure and organelle function of typical plant and animal cells by referring to Figure 6.4 and Figure 6.5. Use the models at your table to help you visualize the organelles and cell structures.

COMMON EUKARYOTIC STRUCTURES AND ORGANELLES

1. **Nucleus:** Large, often centrally-located organelle that stores genetic information in the form of DNA. The nucleus is surrounded by a double membrane called the *nuclear envelope.* While the nuclear envelope is very restrictive, small openings called *nuclear pores* can regulate the passage of certain molecules into or out of the nucleus.

2. **Nucleolus:** Concentrated area within the nucleus that helps make ribosomes.

3. **Cytoplasm:** The semifluid material that fills the interior of a cell and is enclosed by the cell membrane.

4. **Cell Membrane:** The surface made of *phospholipid* molecules (studied in the next lab) that surrounds all cells. The cell membrane is *inside* of the cell wall (if present), and is semipermeable, meaning only certain substances can pass through the membrane. The cell membrane is also known as the *plasma membrane.*

5. **Cell Wall:** A rigid barrier that exists *outside* of the cell membrane in certain cells. In plant cells, the cell wall is made of cellulose and provides support and protection. Animal cells *do not* have a cell wall.

6. **Ribosomes:** Cellular structures that help manufacture proteins. Ribosomes are *not* organelles, and are found in eukaryotes and prokaryotes (see Figure 6.2).

7. **Rough ER:** The rough endoplasmic reticulum is a series of membranous sacs that help to synthesize and modify proteins. The rough ER is distinguished from the smooth ER by the presence of ribosomes, which associate with the rough during protein synthesis.

8. **Smooth ER:** The smooth endoplasmic reticulum is a series of membranous sacs that lack ribosomes and help detoxify harmful compounds and also synthesize lipid molecules.

9. **Golgi Apparatus:** A stack of membranous sacs that package and secrete proteins using vesicles (membrane bubbles).

10. **Lysosome:** A vesicle containing enzymes that hydrolyze (break down) large molecules and cell parts.

11. **Peroxisome:** A vesicle containing enzymes that break down fatty acids into hydrogen peroxide, which can then be used to break down or modify other compounds, including alcohol.

12. **Mitochondrion:** A large organelle that generates energy (in the form of ATP) through cellular respiration. The mitochondrion is surrounded by two membranes.

13. **Chloroplast:** Organelles found in plant cells (not animal cells) that convert solar energy into sugars in the process of *photosynthesis*.

14. **Centrioles:** Cylindrically-shaped organelle that organizes proteins called microtubules. Plant cells *do not* have centrioles.

15. **Central Vacuole:** A large compartment in plant cells that stores water and waste products for the cell.

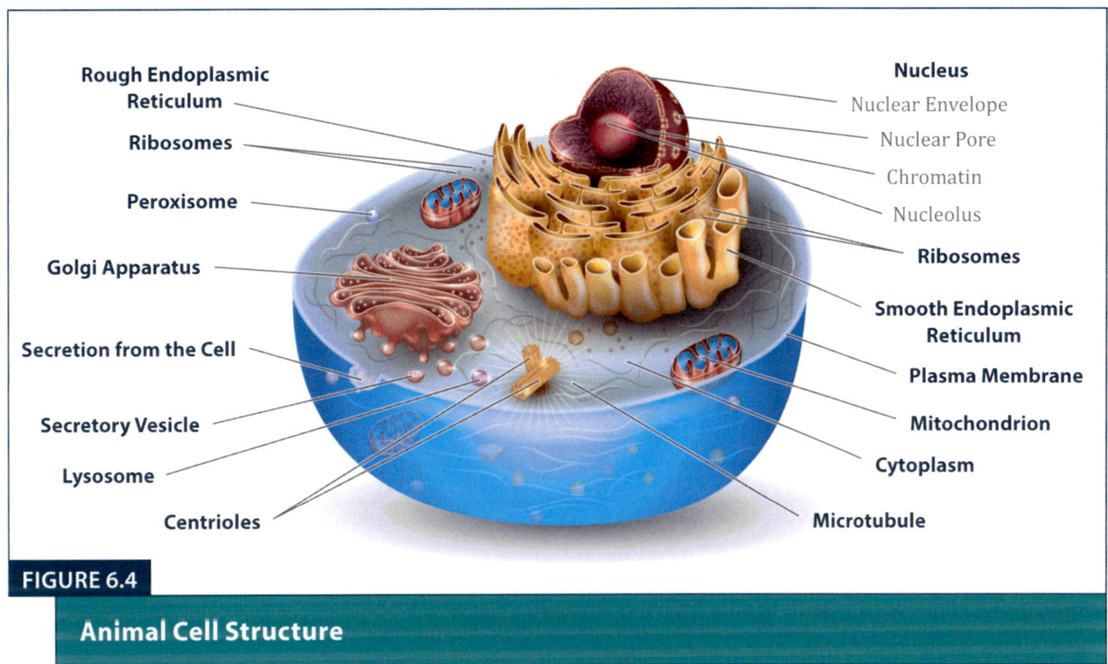

FIGURE 6.4

Animal Cell Structure

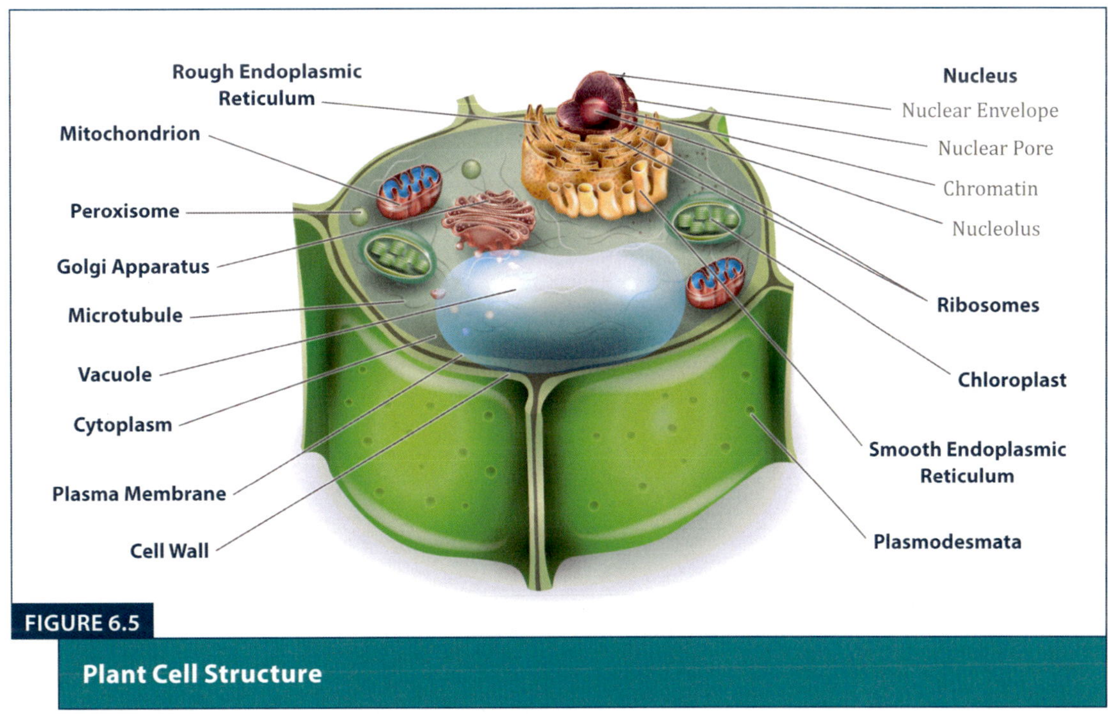

FIGURE 6.5

Plant Cell Structure

PROCEDURE 6.3: OBSERVING PLANT AND ANIMAL CELLS

In this procedure, you will be making preparations of live plant and animal cells. For each sample, you will sketch your specimen at 400× magnification and label any cellular components you can identify.

First, you will be making a wet mount of *Elodea* cells. *Elodea* is a genus of aquatic plant with thin, flat leaves that are easy to view under a microscope. To begin, obtain a clean glass slide and coverslip. Add a drop of water from the beaker containing the plants, and remove a small piece of leaf from the plant. Add the piece of leaf to the drop of water and then add your coverslip. As always, begin observing your specimen at 4× power before advancing to 10× and 40×. Sketch a few cells from your specimen in the circle to the right, and label any of the cellular components from the list of structures and organelles from the previous procedure.

For your next specimen, you will observe animal cells by sampling from your own tissues. Obtain a clean, glass slide and coverslip and a toothpick from the container at your table. Because the nuclei of animal cells are transparent, they will not show up on a slide without being stained. Add a drop of *methylene blue* dye from your bin to the center of your slide. Using the toothpick, gently scrape the inside of one of your cheeks and then swirl the toothpick in the drop of dye to transfer the cells into the dye. Apply the coverslip and dispose of the toothpick in the **hazardous waste container** (ask your instructor if you are unsure of the location of this bin). Use scanning power and low power to find cells within your sample, then proceed to high power to view them in greater detail. Sketch a few of your cells in the circle to the right and answer the *thinking critically* question that follows.

Elodea Cells at 400× Magnification

Human Cheek Cells at
400× Magnification

Thinking Critically: Cell Walls

Plants, and fungi (eukaryotes) and bacteria (prokaryotes) typically have cell walls surrounding their cells. Animals, as we have seen, never have cell walls. Think about what makes animals unique (especially compared to plants and fungi), and explain one reason why it would *not* be advantageous for an animal cell to contain a cell wall.

PROCEDURE 6.4: SAMPLING FOR SINGLE-CELLED EUKARYOTES IN POND WATER

As you now know, eukaryotic cells are large and complex. Many eukaryotic organisms are composed of only a single cell, and are thus called unicellular organisms. These organisms are often aquatic, and can be found readily in freshwater ecosystems. Biologists used to lump these organisms into a taxonomic (classification) group called the "protists," similar to how plants, animals, and fungi are grouped separately. The problem is that not all unicellular eukaryotes are closely related, and many are as distantly related as you are to a plant. Figure 6.6 displays some common unicellular eukaryotes that can be found in pond water. Prepare a wet-mount sample using a drop or two of the water from the pond water bucket on the side bench.

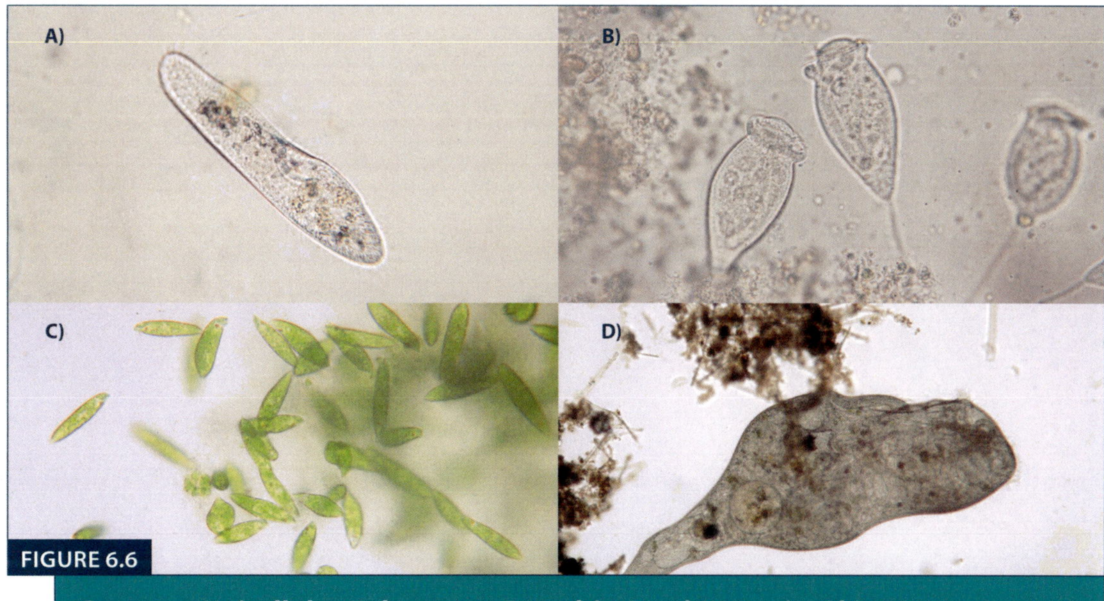

FIGURE 6.6

Common Unicellular Eukaryotes Found in Pond Water in Ohio. A) Paramecium, B) Vorticella, C) Euglena, and D) Stentor

Which of the specimens from Figure 6.6 were you able to see under your microscope?

With the help of your instructor, what additional specimens can you identify?

Describe the type(s) of movement that your specimen(s) exhibit within the water.

MEMBRANE TRANSPORT

OBJECTIVES

By the end of this lab exercise, students should be able to

- describe the structure of the plasma membrane;
- describe how membranes are selectively permeable;
- describe how diffusion occurs in cells;
- analyze and explain the results from experiments simulating diffusion in cells;
- distinguish between hypertonic and hypotonic solutions; and
- define **diffusion, plasmolysis,** and **contractile vacuole.**

INTRODUCTION

In previous lab activities, you have discovered the complex nature of cells, including some of the components that allow cells to function. Most every cell depends upon the ability of certain substances to enter or exit the cell. All cells are surrounded by a *plasma membrane,* which surrounds the cytoplasm and helps regulate the passage of specific substances into and out of the cell. The plasma membrane (sometimes simply called the cell membrane) is *selectively permeable,* meaning that only certain substances can

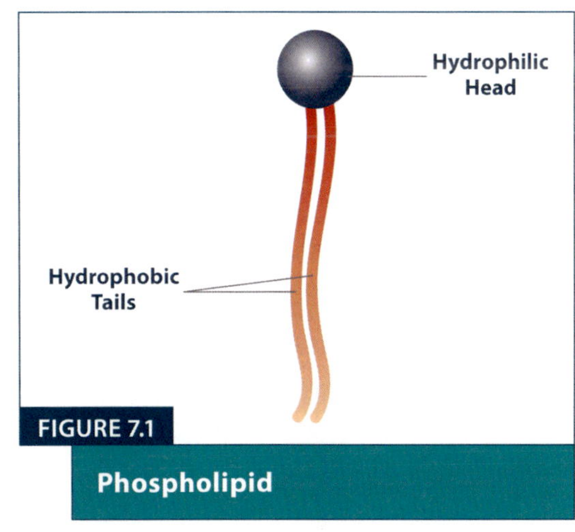

pass freely through it. This property of membranes is due in part to the chemical nature of their main constituent, the phospholipid. Phospholipids are molecules that have polar "head" regions and nonpolar "tails" (see Figure 7.1). The heads, which contain a phosphate group, are hydrophilic, and point toward water. The tails, which are hydrophobic, point away from water. Plasma membranes form as a bilayer of phospholipids (see Figure 7.2)

The hydrophobic layer of a plasma membrane prevents most polar molecules, including proteins, from crossing freely. Lipids, which are nonpolar molecules, can typically cross the plasma membrane relatively easily. Certain small polar molecules, like water, can cross the membrane, but do so slowly. Ions, which are charged particles, cannot freely cross the membrane. When a substance *can* pass through the plasma membrane, we can predict in which direction it will travel (into or out of a cell) based on the concept of diffusion, which we will explore in this lab.

Imagine that you are ready to check out at the grocery store, and only one checkout lane is open. The lane has nine people waiting, so you reluctantly proceed to the back of the line. Suddenly, a second lane opens up, and people from your lane begin to move to the new lane. Without doing any communicating, we can safely assume that about half of the people from the old lane would move to the new lane, rather than one, or all ten. The movement of people would stop when a balance was reached. A similar principle applies to substances that diffuse across membranes. **Diffusion** is the movement of a substance from a higher concentration to a lower concentration. Net diffusion typically ceases when a balance, or equilibrium is reached on both sides of the membrane (see Figure 7.3).

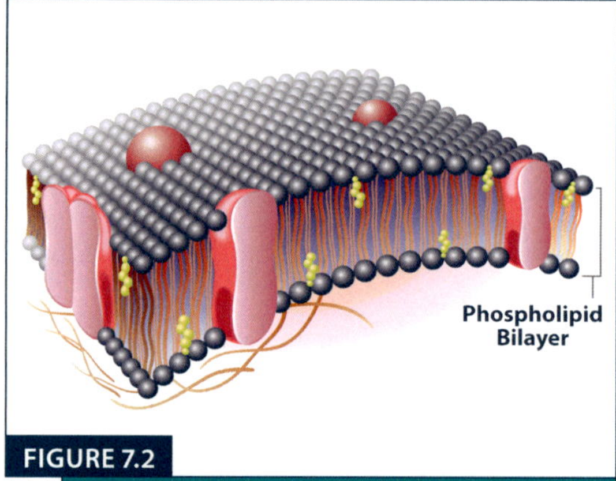

FIGURE 7.2

Plasma Membrane. The plasma membrane forms as a bilayer, with hydrophobic tails pointing away from water inside and outside of a cell.

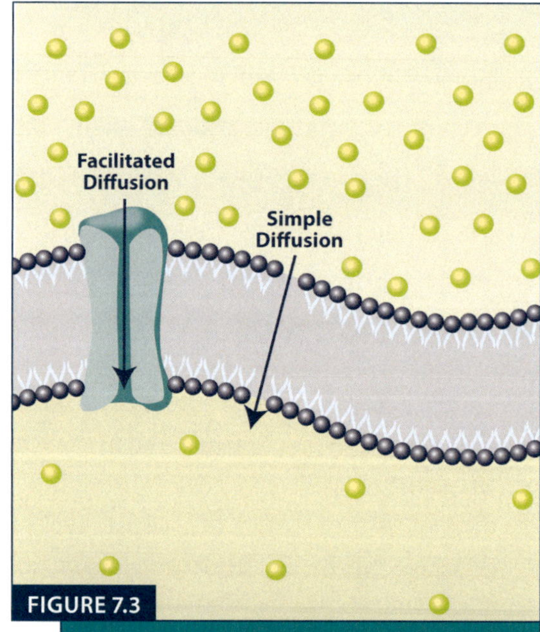

FIGURE 7.3

Diffusion. Diffusion is the movement of a substance from an area of higher concentration to an area of lower concentration.

PROCEDURE 7.1: SIMULATING DIFFUSION ACROSS A MEMBRANE

In this procedure, you will create a model cell using dialysis tubing. Dialysis tubing, which is used to filter blood in people with kidney disorders, is like a plasma membrane in that is permeable to some substances and not others.

To begin, take your dialysis tubing and open one end by rubbing the tubing between your thumb and index finger (as though you were trying to open a garbage bag). Seal one end of the tubing using one of the plastic clips at your table, and then fill the bag about halfway with a solution of glucose and starch from the bottle at your table. Seal the other end of the dialysis tubing and fill a 250-mL beaker filled with about 200 mL of distilled water. Add about 10–15 drops of iodine solution to the beaker, or until the water turns an amber color. Once you have added the iodine, place your sealed dialysis tubing into the beaker and leave the fluid portion fully submerged for 40–50 minutes. Your completed set-up should resemble Figure 7.4.

FIGURE 7.4

Simulating a Cell Membrane. Dialysis tubing can simulate a cell membrane when submerged in specific solutions.

As you begin timing, work with your lab partners to develop a hypothesis as to how diffusion will occur throughout the experiment. Be sure to mention which compounds you think will diffuse and which will not, and include why or why not. After the bag has sat for the allotted time, proceed to the conclusions section.

HYPOTHESIS: SIMULATING DIFFUSION ACROSS A MEMBRANE

CONCLUSIONS: SIMULATING DIFFUSION ACROSS A MEMBRANE

In order to determine if iodine or starch diffused, remember from previous experiments that iodine will turn a dark purple shade when mixed with starch.

Did iodine diffuse into the bag? ______________________

How do you know? ______________________

Did starch diffuse out of the bag? ______________________

How do you know? ______________________

Why do you suppose the starch was unable to diffuse?

____

____

____

To determine if glucose diffused out of the bag, we can do a *Benedict's Reagent* test using test tubes. Obtain two clean test tubes and two clean pipette tips. Transfer 2 mL of fluid *from inside the dialysis tubing* into one of the test tubes. Using a the second tip, transfer 2 mL of fluid from the center of the beaker into the second test tube. Add 5 drops of Benedict's Reagent to each tube and place the tubes in boiling water for one minute. Compare the colors of the solutions in each tube to Table 4.3 from Experiment 4.

What color is the solution in the test tube that was taken from the dialysis tubing?

____

What color is the solution in the test tube that was taken from the beaker?

____

Did glucose diffuse out of the bag? ______________________

PROCEDURE 7.2: MEASURING THE RATE OF DIFFUSION ———

Now that you understand the role of concentration in diffusion, you might begin to wonder what can impact the rate at which a substance will diffuse. In this procedure, you will use three different pieces of dialysis tubing to represent plasma membranes. Each section of tubing will contain a different concentration of NaCl solution, and will be placed in a beaker containing distilled water. Because the bags will each contain a solution where water is *less* concentrated in the bag than outside the bag, we can expect water to diffuse into the bags. Your independent variable is thus the concentration of NaCl in each dialysis bag. You will be testing the effect of a *concentration gradient* (how concentrated something is on either side of a membrane) on the rate of diffusion.

Begin by opening three new pieces of dialysis tubing. Clip one end of each piece of tubing, and add three different concentrations of NaCl solution (see Table 7.1) to fill each bag halfway up with a different solution. Clip the other end of the bags, and then take a mass measurement of each bag. Fill in your mass measurements (taken in grams) for each dialysis bag in Table 7.1.

Measuring the Rate of Diffusion

TABLE 7.1				
BAG CONCENTRATION	**INITIAL MASS**	**MASS AT 10 MINUTES**	**MASS AT 20 MINUTES**	**MASS AT 30 MINUTES**
1% NaCl				
5% NaCl				
10% NaCl				

Once you have taken your initial mass readings, place each bag in a separate 250-mL beaker filled with 200 mL of distilled water. You will take mass readings of each bag at ten minute intervals, for a total of 30 minutes. The bag that gains the most mass will be the bag that experienced the most rapid diffusion of water. As you begin timing, work with your partners to hypothesize how the rate of diffusion will differ in each bag.

HYPOTHESIS: MEASURING THE RATE OF DIFFUSION

CONCLUSIONS: SIMULATING DIFFUSION ACROSS A MEMBRANE

Based on your data from Table 7.1, what can you conclude about the relationship between the concentration of salt in a solution and the rate at which water diffuses in?

PROCEDURE 7.3: OBSERVING DIFFUSION IN PLANT CELLS

Plants, like all organisms, are dependent upon water to live. If you have ever grown house plants, you may be familiar with the characteristic wilted appearance of a plant that is dehydrated. Some plants, like the *Elodea* plants we have used in lab before, are adapted to live in water. We can begin to see signs of dehydration in *Elodea* cells long before the leaves begin to wilt by using a microscope.

All cells contain water, but sometimes a cell can be in an environment where water diffuses into or out of the cell. If a cell is in a *hypotonic* solution, the water in the solution is more concentrated than the water in the cell, so the cell gains water. In a *hypertonic* solution, a cell will lose water because the water inside the cell is more concentrated than it is in the outside solution.

FIGURE 7.5

Plant Cells That Have Undergone Plasmolysis

A plant will undergo a process called **plasmolysis** when in a hypertonic solution. During plasmolysis, the plasma membrane pulls away from the cell wall as the cell loses water (see Figure 7.5). In this experiment, you will expose *Elodea* cells to hypertonic and hypotonic solutions, and visualize the effects under a microscope.

To begin, obtain two clean microscope slides and two coverslips. Onto one slide, add a drop or two of distilled water. This sample will represent a hypotonic solution. To the second slide, add a drop or two of 10% NaCl (table salt) solution. This will represent a hypertonic solution. Remove two *Elodea* leaves from the beaker and place one leaf onto each slide. Add a coverslip and begin viewing at scanning power. Proceed to low- and high-power magnifications.

CONCLUSIONS: OBSERVING DIFFUSION IN PLANT CELLS

Were the cells gaining or losing water in the distilled water slide? _______________

Describe how the cells in the hypotonic solution appeared that led you to conclude that water diffused into the cells.

Were the plants gaining or losing water in the NaCl solution slide? _______________

Describe how you were able to tell that *plasmolysis* occurred.

Thinking Critically

When plant cells are in a hypotonic solution, they fill with water, which causes the plasma membrane to expand and push against the cell wall. This is called *turgor pressure*. Why do you think it would be advantageous for a plant to produce pressure within its cells? Work with your partners to explain how turgor pressure might benefit plants.

PROCEDURE 7.4: OBSERVING DIFFUSION IN PARAMECIUM CELLS

A paramecium is a type of single-celled eukaryote that lives in aquatic environments. They have a "slipper" shape (see Figure 7.6), and move by beating cilia, which are hair-like structures made of proteins called microtubules. When paramecia live in freshwater systems like lakes and ponds, they are constantly surrounded by water that is more concentrated than the water inside of them. Similar to *Elodea,* this means that they are constantly taking in water. Paramecium use specialized cellular compartments called **contractile vacuoles** to help them cope with a hypotonic environment. A contractile vacuole is an organelle that fills with water inside the paramecium and then expels the water by dumping it out of the plasma membrane.

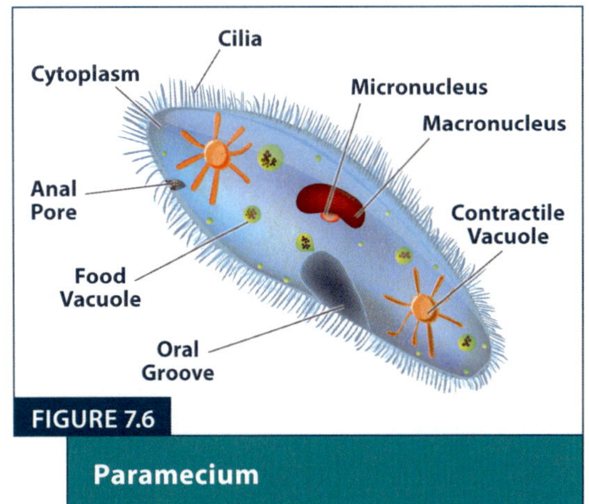

FIGURE 7.6

Paramecium

Prepare a wet-mount sample by obtaining a clean glass slide and coverslip. Place a drop or two of solution from the paramecium culture vial onto the center of your slide (your instructor may ask you to use a separate solution to slow the cells down if available). Add the coverslip and begin viewing at scanning power. Scan the field of view until you find a paramecium, and then advance to low and high powered objectives.

At high power, observe the tiny circles inside of the paramecium that seem to appear and disappear in a fraction of a second. These are contractile vacuoles, which are actively filling with water and merging with the plasma membrane to remove the water from the cell.

Sketch your paramecium at high power below, and indicate where the contractile vacuoles can be seen.

Thinking Critically: Observing Diffusion in Paramecium Cells

Both *Elodea* and *Paramecium* live in hypotonic environments, but *Elodea* do not need to expel water in the same way the a paramecium does. Think about what you have learned about plant cell structures, and explain why you think plant cells do not need to expel water in hypotonic environments.

8

CELLULAR RESPIRATION AND FERMENTATION

OBJECTIVES

By the end of this lab exercise, students should be able to

- ⊙ provide the overall equation for cellular respiration;
- ⊙ distinguish between aerobic and anaerobic respiration;
- ⊙ define **energy, ATP, cellular respiration, glycolysis,** and **fermentation;** and
- ⊙ design, conduct, and analyze an experiment on fermentation rates in yeast cells.

INTRODUCTION

You have probably been told at some point in your life that it is important to eat break-fast in order to get the energy you need to function at school or work, but you might not have thought about the specifics of how it is that your body can convert the food that you eat into usable energy. In physics, **energy** is often defined as the capacity to cause change (or do work). Some reactions, including many of those that take place in cells, require continuous energy input to occur. Other reactions release energy as they occur. Many cellular reactions involve building complex structures from smaller subunits (a process called anabolism) or breaking down complex structures into sim-pler subunits (catabolism). Most every cell on the planet uses the catabolism of **ATP** to power reactions (see Figure 8.1). ATP (Adenosine triphosphate) is the cell's main energy-carrying molecule, and is often called "the energy currency" of cells because it is generated and used up in a way similar to how money can be made and spent.

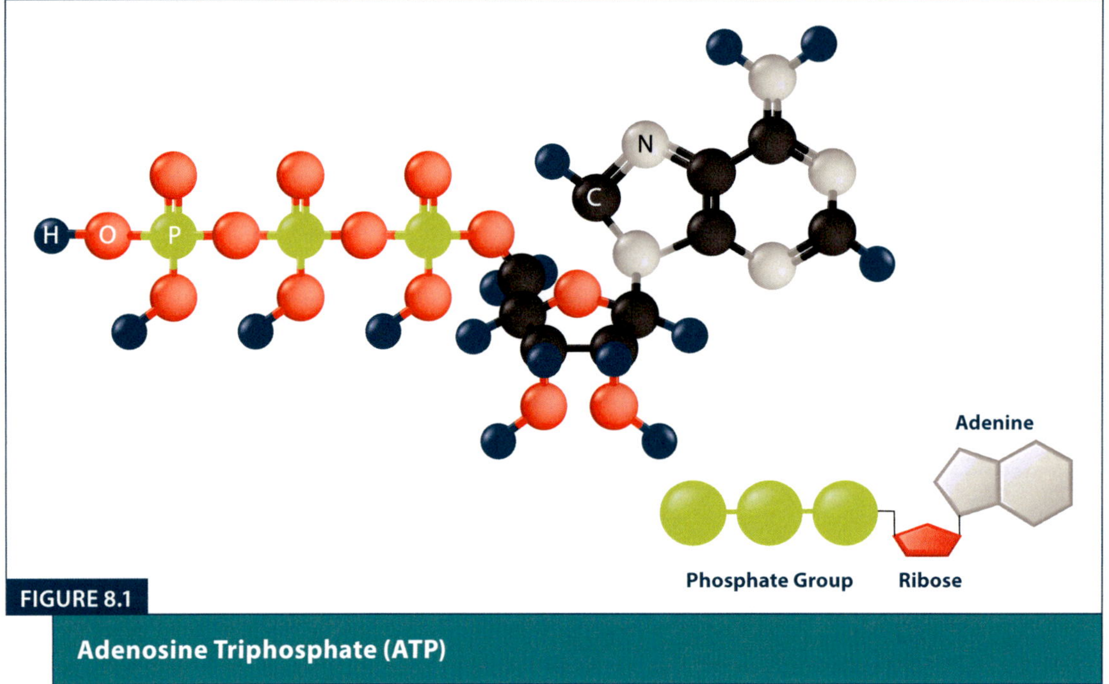

FIGURE 8.1

Adenosine Triphosphate (ATP)

In cells, it takes the energy of another reaction to build ATP. For many cells, glucose is broken down in a series of reactions to release energy that is used to build ATP. The entire process of converting nutrient molecules (like glucose) into energy and waste products is called cellular respiration. The overall equation for aerobic **cellular respiration** can be seen in Eq 8.1.

$$C_6H_{12}O_6 + 6O_2 \longrightarrow 6CO_2 + 6H_2O + ATP \qquad \text{EQ 8.1}$$

Glycolysis is the first step in breaking down a glucose molecule. In this process, glucose (a six-carbon molecule) is broken down into two three-carbon *pyruvate* molecules (see Figure 8.2).

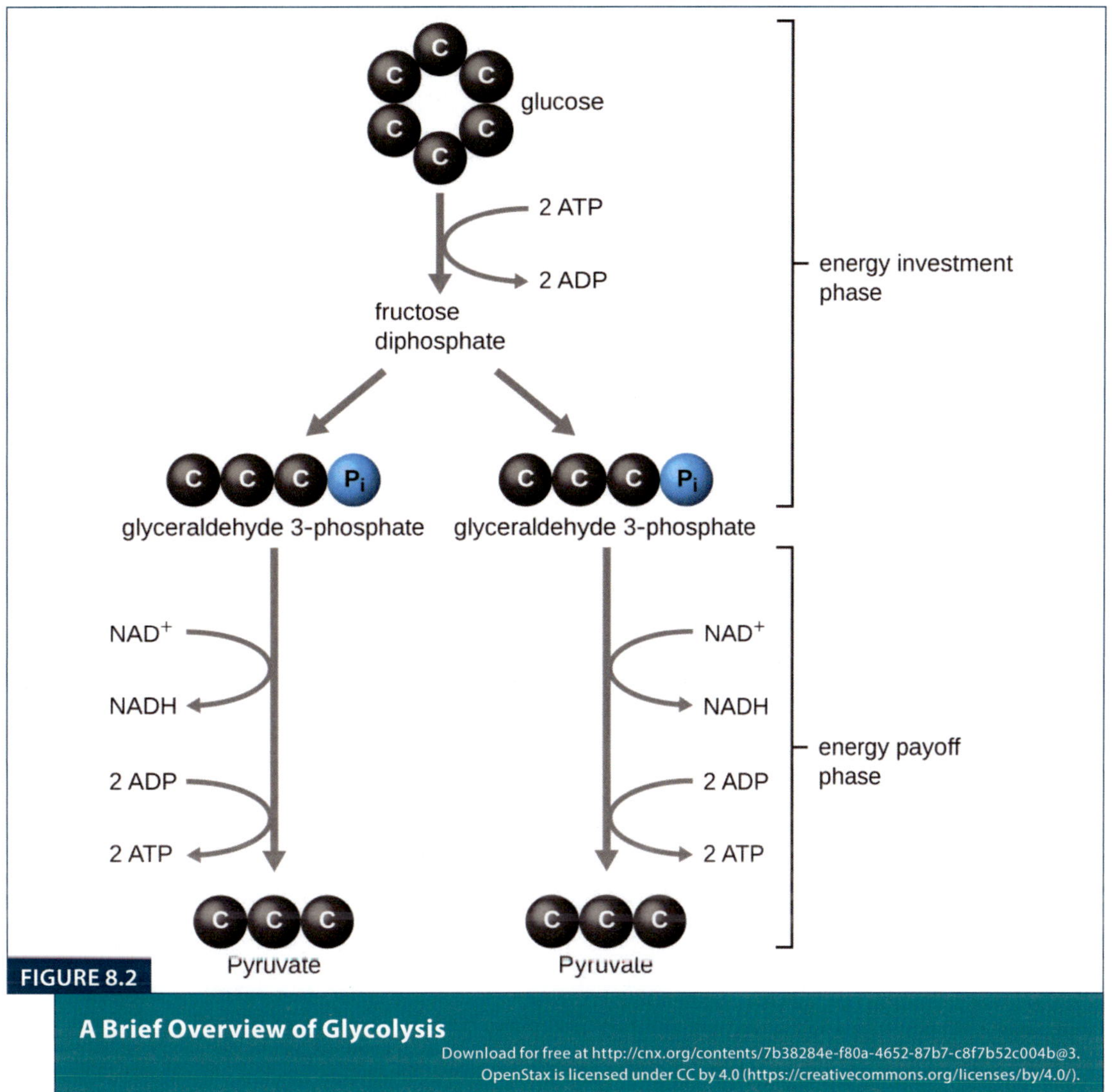

FIGURE 8.2

A Brief Overview of Glycolysis

What occurs after glycolysis differs depending on cell type. In most eukaryotic cells, the pyruvate molecules can be sent to the mitochondria, where they are modified in a series of reactions that produces many more ATP molecules. This process is called *aerobic respiration,* as it requires oxygen. In cells without mitochondria (and in some eukaryotic cells including yeast cells and muscle cells), the pyruvate from glycolysis can be converted into a waste product in a process called fermentation. Ethanol fermentation, for example, begins with glycolysis and ends with the production of carbon dioxide, ATP, and ethanol (see Eq 8.2). It is important to remember that during fermentation, ATP is only produced in glycolysis. Converting pyruvate to ethanol does not provide any usable energy, but does restore a molecule (NAD^+) that can help cells go through glycolysis again to make more ATP. The overall reaction for fermentation is as follows:

$$C_6H_{12}O_6 \longrightarrow 2CO_2 + 2C_2H_5OH + 2ATP \qquad \text{EQ 8.2}$$

Notice that Eq 8.2 does not include oxygen as a reactant. Fermentation is an *anaerobic* process, meaning it does not require oxygen to occur. In this lab, we will be measuring fermentation in yeast cells, which are frequently used in the food industry to make products for human consumption (Figure 8.3).

FIGURE 8.3

Products of Yeast Fermentation. Products like bread and beer are made using yeast fermentation.

PROCEDURE 8.1: MEASURING FERMENTATION IN YEAST: DESIGNING AN EXPERIMENT

In this experiment, you will be measuring the rate of fermentation in yeast cells by mixing live yeast with sugar (sucrose) and water. As the yeast cells digest the sugar, they will produce carbon dioxide (CO_2) as a waste product. The CO_2 can be trapped using a rubber or nitrile glove. We will be measuring fermentation as a function of the expansion of the glove as it fills with CO_2 from the yeast (Figure 8.4). You must now work with your lab partners to design an experiment that will measure the influence of an independent variable (treatment) of your choosing on fermentation.

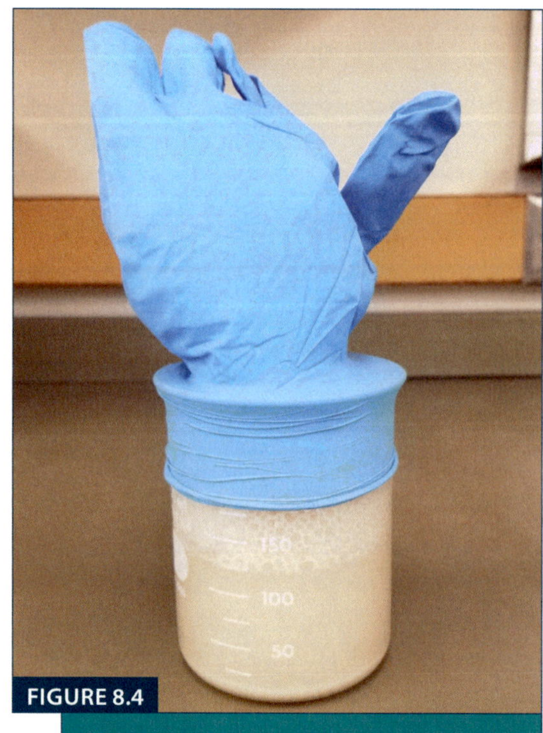

FIGURE 8.4

Yeast Fermenting in a Beaker Sealed with a Glove

You will have four beakers to use in your experiment. Each will have some combination of yeast, sugar, and water, and will be placed in a specific temperature to react for **50 minutes.** One beaker will be a positive control, which will give you a standard of comparison for your other trials. In this beaker, you will place the following contents:

Beaker 1 (control) Contents: **15 g sucrose, 3 g yeast, 250 mL distilled water.** This beaker will be placed in a warm water bath at **40 °C.**

Work with your lab partners to design your experiment, and be sure to include your independent variable. Remember that only *one* independent variable should be tested. Decide how you want to alter your other three beakers and outline your experiment below. Include your *hypothesis* with your experimental outline.

EXPERIMENTAL OUTLINE: MEASURING YEAST FERMENTATION

Independent Variable: ______________________________________

Beaker 2 Contents: ____

Beaker 3 Contents: ____

Beaker 4 Contents: ____

HYPOTHESIS

Thinking Critically: Measuring Yeast Fermentation

Yeast cells are fungi, and fungi are eukaryotes like animals and plants. Yeast thus have mitochondria, and are capable of using *aerobic respiration* like us to make much more ATP (the entire aerobic respiration cycle, beginning with glycolysis, generates around 30–32 ATP molecules). While this route produces more ATP per glucose molecule, it also takes more time. Working with your lab partners, try to come up with *two* conditions that would cause yeast cells to go through anaerobic fermentation instead of aerobic respiration.

PROCEDURE 8.2: MEASURING FERMENTATION IN YEAST: CONDUCTING AN EXPERIMENT

Now that you have designed your experiment and have formulated a hypothesis, you can proceed to set up your experiment using the following instructions:

1. Fill your beakers with distilled water to the 250-mL mark (unless you are manipulating the amount of water as your independent variable).

2. Using an electronic balance, measure the quantities of sugar you will be adding to your beakers. If you are not manipulating sugar as your independent variable, this quantity should be 15.0 g per beaker. Add the sugar to each beaker and stir using a stirring rod.

3. Using an electronic balance, measure the quantities of yeast you will be adding to your beakers. If you are not manipulating yeast as your independent variable, this should be 3.0 g per beaker. Add the yeast to each beaker and stir using a stirring rod. Each beaker should be of similar consistency before incubating (ask your instructor to check the consistency before incubating).

4. Attach a nitrile glove to each beaker (make sure the gloves are placed similarly on each beaker) and place each beaker in the warm water bath (incubator) at 50 °C. If you are manipulating temperature, place each beaker in its appropriate temperature. Allow beakers to incubate for 50 minutes.

5. At each *ten-minute interval,* measure the volume of gas in your glove (in cm) using a ruler. To do this, measure the *highest point of inflation* in the glove (see Figure 8.5). Your instructor will demonstrate how to do this correctly.

6. Keep a record of the gas levels in each glove by entering the data at each interval in Data Table 8.1.

7. When the experiment is complete, bring your beakers to the sink and carefully remove the gloves. Dump the contents down the drain and dispose of the gloves.

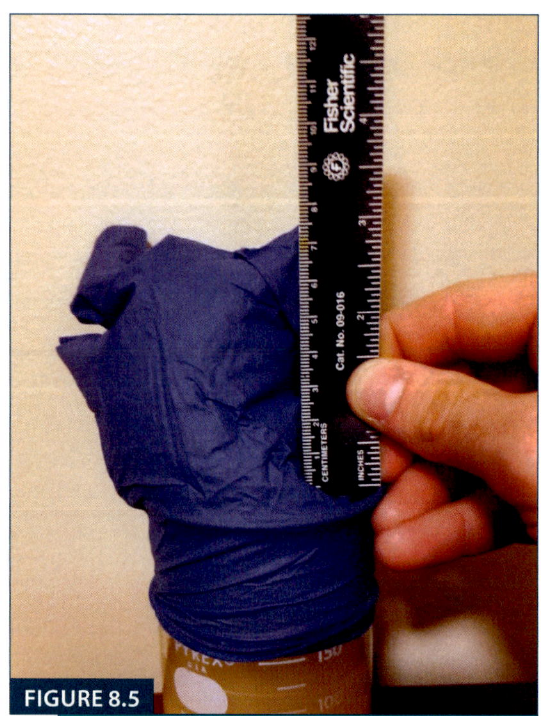

FIGURE 8.5

Measuring Gas Production in a Rubber Glove

TABLE 8.1

BEAKER	1	2	3	4
Gas Height (cm) 10 Minutes				
Gas Height (cm) 20 Minutes				
Gas Height (cm) 30 Minutes				
Gas Height (cm) 40 Minutes				
Gas Height (cm) 50 Minutes				

CONCLUSIONS: MEASURING YEAST FERMENTATION

Briefly describe the results of your experiment, including any observable effects that your independent variable had on the rate of fermentation in your yeast samples.

If possible, work with your lab partners to describe *why* you think your independent variable had the effect (or lack of effect) that you observed.

Was your initial hypothesis supported by your data? ______________________________

PHOTOSYNTHESIS

OBJECTIVES

By the end of this lab exercise, students should be able to

⊙ explain the role of photosynthesis in food webs;

⊙ provide the overall equation for photosynthesis and compare the products and reactants to those of cell respiration;

⊙ describe the major components of a chloroplast;

⊙ explain the role of plant pigments in photosynthesis and the science of chromatography;

⊙ distinguish between the function and products of the light reaction and Calvin Cycle;

⊙ conduct and analyze an experiment that measures photosynthesis in plants; and

⊙ define **producer, consumer, chlorophyll, net photosynthesis,** and **gross photosynthesis.**

INTRODUCTION

In the previous lab, you learned of the important role of glucose in generating ATP for energy in cells. Virtually every organism on the planet utilizes glycolysis to break down glucose for energy, including animals like owls and cats that consume only meat (their cells undergo processes that can convert other organic molecules into glucose). In order for animals, fungi, and many other eukaryotic (and prokaryotic) organisms to obtain glucose, they must consume or feed on other organisms. A **consumer** is an organism that feeds off of the energy of another organism. Since consumers are not capable of making their own food, they must consume, or feed. A **producer,** like plants, algae, and cyanobacteria, can produce organic "food" molecules like glucose. The importance of producers in biology can be seen in a *food web,* which is a representation of "what-eats-what" in an ecosystem (see Figure 9.1). While sometimes far-removed from a producer in a food web, consumers are always reliant upon the high-energy molecules, like glucose, that are made by producers in the web.

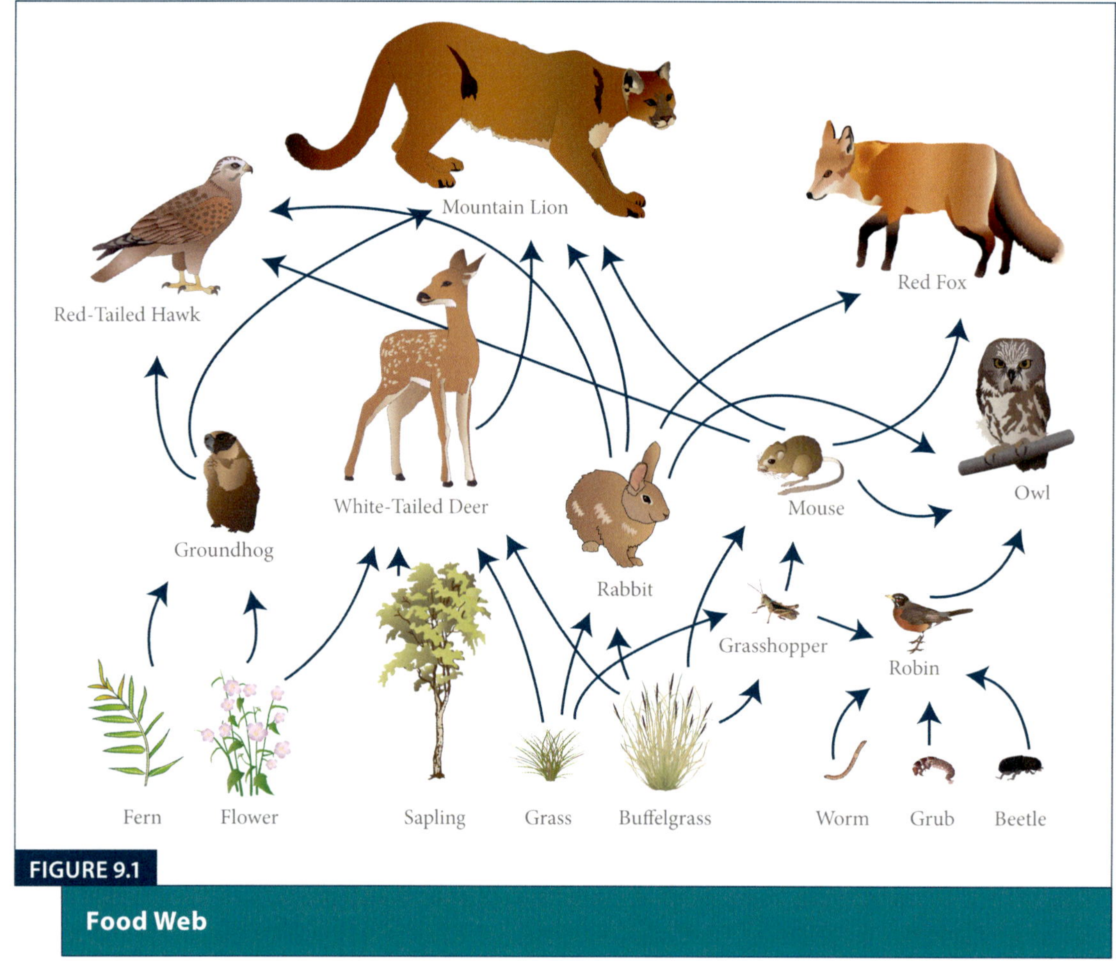

FIGURE 9.1

Food Web

Photosynthesis is a process in which the sun's energy is used to make carbohydrates. In addition to solar energy, the reaction requires carbon dioxide (CO_2) and water (H_2O). The major products of the reaction are glucose ($C_6H_{12}O_6$) and oxygen (O_2). The overall reaction can be summarized as

$$6CO_2 + 6H_2O + \text{Solar Energy} \longrightarrow C_6H_{12}O_6 + 6O_2.$$

While photosynthesis involves a series of complex biochemical reactions, we can divide the overall process into two main reactions: the *Light Reaction* and the *Calvin Cycle.* In the Light Reaction (sometimes called the Light-Dependent Reaction), light is converted into chemical energy. In the Calvin Cycle (sometimes called the Light-Independent Reaction), the energy generated in the Light Reaction is used to power the building of sugars like glucose. In plants, both reactions, which together make up the process of photosynthesis, occur in organelles called chloroplasts (see Figure 9.2), which you have studied in previous labs.

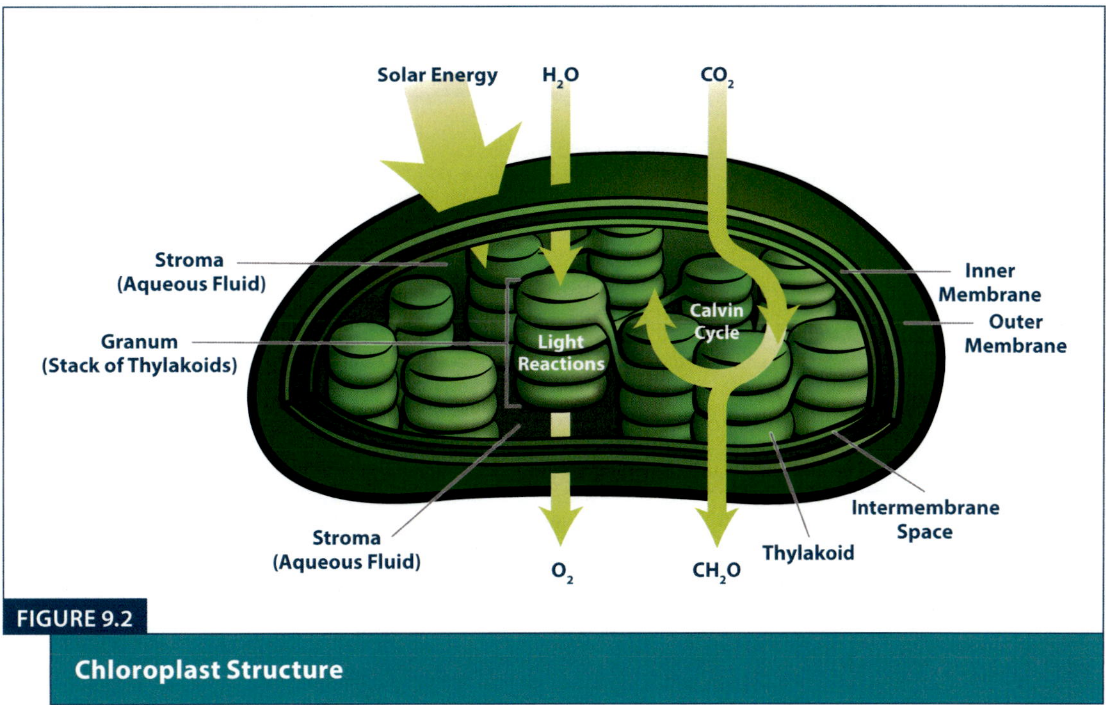

FIGURE 9.2

Chloroplast Structure

Like mitochondria, chloroplasts contain two membranes. The fluid portion of the chloroplast that fills the interior of the organelle is called the **stroma.** The Calvin Cycle takes place in the stroma of chloroplasts. Thylakoids are disk-shaped structures that contain chlorophyll pigments in their membranes. A stack of thylakoids is called a **granum.**

The two procedures of this lab will examine aspects of the Light Reaction, which occurs in the thylakoids of a chloroplast. Remember that the Light Reaction involves generating chemical energy from sunlight.

PROCEDURE 9.1: CHROMATOGRAPHY

In Ohio, the autumn season is associated with tree leaves changing from green to various shades of red, orange, yellow, and brown. The color of a leaf is determined by pigments, which are molecules that appear a certain color because they *absorb* or *reflect* particular components of light. Some of these pigments play a role in photosynthesis, and others promote cell health and metabolism in other ways. In living things, pigments are typically proteins, which are produced by the cell. The visible light that shines from the sun (or a light bulb), often called white light, contains multiple colors (wavelengths). These colors can be depicted in the form of a spectrum (see Figure 9.3). When pigments reflect a color of light, that means the color is not being absorbed. For example, if you see a cardinal in a tree, the red coloration that you see is due to pigments in the bird's feathers that reflect red light, which reaches your eye. Objects that appear white contain molecules that reflect all of the colors of light, while objects that appear black contain molecules that absorb all of the light that shines on them and reflect no light. You may have been told by your parents to avoid wearing black clothing on a sunny day because black materials absorb so much of the sun's light.

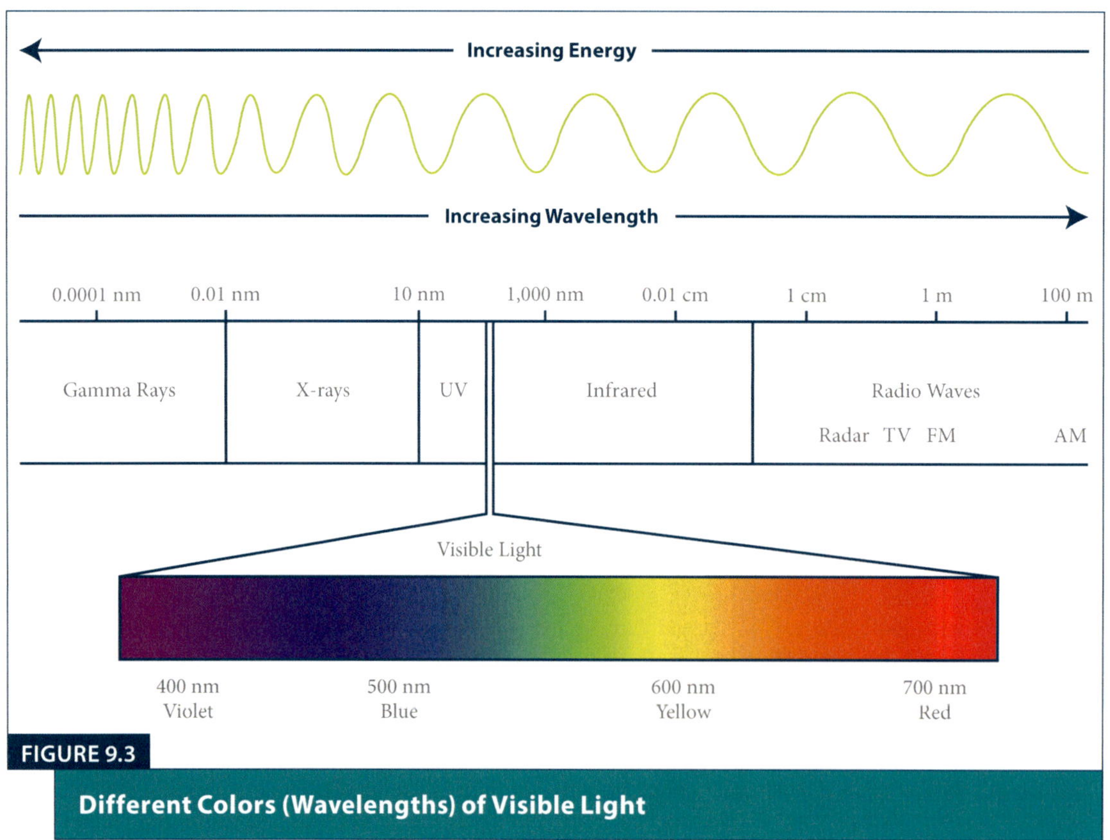

FIGURE 9.3

Different Colors (Wavelengths) of Visible Light

When leaves change colors in the autumn, it is not because they are producing new pigment molecules that they were not producing in the spring and summer. Instead, pigments that they always produced are no longer masked by **chlorophyll** pigments, which are pigments that reflect green light and take part in photosynthesis. Plants thus always contain multiple pigment molecules, which can reflect different types of light and give a leaf a specific color depending upon how abundant the molecules are in the leaf. In this procedure we will use a process called *chromatography*, which separates components of a solution based on their solubility in a specific solvent. Follow the instructions below to separate the pigments found in spinach leaves.

1. Obtain a 200-mL Erlenmeyer flask with a cork (the cork will have a hook attached to it).

2. Attach a strip of chromatography paper to the hook so that the bottom of the strip barely touches the bottom of the flask (see Figure 9.4).

3. Measure 1 cm from the bottom of the strip and place a small dot in the middle of the strip with a pencil (*do not use a pen*).

4. Under the fume hood, add 10 mL of chromatography solution to your Erlenmeyer flask.

5. Place the strip of chromatography paper on a paper towel. Using a capillary tube (see Figure 9.5), apply a small drop of the plant leaf solution to the pencil dot on the paper strip.

FIGURE 9.4

Attaching Chromatography Paper. Your chromatography paper strip should just barely touch the bottom of your flask.

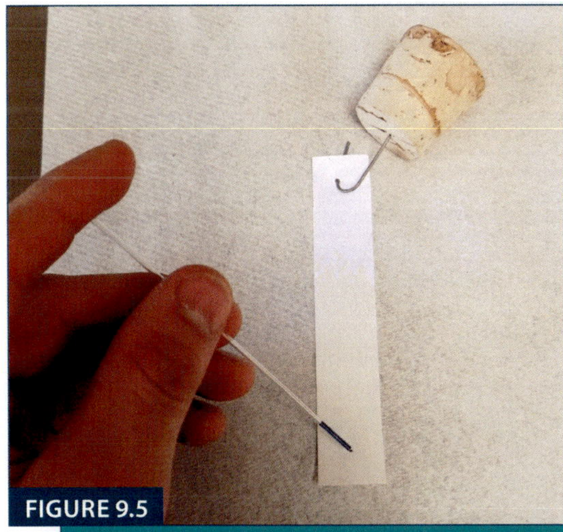

FIGURE 9.5

Using a Capillary Tube. Add plant pigments to your paper strip using a capillary tube.

6. Attach your strip of chromatography paper back on the hook of the cork. Under the fume hood, place the cork in the flask so that the tip of the paper strip contacts the chromatography solution.

7. Bring your assembled flask back to your table and allow the solution to soak the paper until the you can see the solution travel to about 1 cm from the upper edge of the paper.

8. Remove the paper strip and place the cork back in the flask. Place the paper strip on a paper towel and mark (with a pencil) where the solution stopped on the strip. Bring your paper strip to the fume hood and let it dry for about five minutes. Observe the color pigments that have separate on the paper, and enter your results in Table 9.1. The pigments often appear in the order given in Table 9.1 (from top to bottom on the chromatogram).

Pigments in Spinach Leaves

TABLE 9.1	
PIGMENT TYPE	**PRESENCE IN CHROMATOGRAM (YES/NO)**
Carotenes (Appears Yellow)	
Xanthophylls (Appears Orange)	
Chlorophyll A (Appears Blue/Green)	
Chlorophyll B (Appears Green)	

Now complete the procedure a second time using leaves from a Japanese maple tree, which appear red even during spring and summer months. Follow the same instructions as with the spinach leaf sample, and complete Table 9.2 after your second chromatogram has dried.

Pigments in Japanese Maple Leaves

TABLE 9.2	
PIGMENT TYPE	**PRESENCE IN CHROMATOGRAM (YES/NO)**
Carotenes (Appears Yellow)	
Xanthophylls (Appears Orange)	
Chlorophyll A (Appears Blue/Green)	
Chlorophyll B (Appears Green)	

RESULTS: CHROMATOGRAPHY

Briefly summarize any observations you can make between the pigment content/separation in the two types of plant leaves.

Thinking Critically

Japanese maple trees appear red during the entire growing season because they usually do not produce as much chlorophyll pigment as other plants, like spinach. If you had a Japanese maple tree, what do you think you could do to the plant in order to cause it to produce more chlorophyll and change the colors of its leaves? Discuss with your lab group and write your response below.

PROCEDURE 9.2: MEASURING THE LIGHT REACTION

In this procedure you will set up an apparatus that allows you to directly measure the products of the light reaction, which will give you an idea of the level of photosynthetic activity of your plants. To do this, you will use a volumeter, which contains a test tube and a rubber stopper. The rubber stopper contains a thin tube that is used to measure volume changes. Because the Light Reaction produces oxygen, we can measure the reaction in plants by watching fluid levels rise in our volumeter (when oxygen is produced, it will cause the fluid level in the volumeter to rise).

Follow the instructions below to measure the Light Reaction in *Elodea* plants.

1. Obtain three pieces of *Elodea* plant from one of the beakers on the side bench. Each piece should be about 7–8 cm in length. Place the plants in your large test tube.

2. Add *sodium bicarbonate solution* to your test tube to cover the plants. You should add enough solution that, when you firmly place the rubber stopper into the test tube, the thin tube extending below the stopper should be barely submerged in the fluid (see Figure 9.6).

3. Place the volumeter in your test tube rack, and use a wax pencil to mark the initial fluid level that you can read in the thin stopper tube (see Figure 9.7).

4. Turn on your lamp and place the lamp about 10–15 cm away from your volumeter. *Make sure that your fluid level in the thin stopper tube begins to*

FIGURE 9.6

A Volumeter Containing *Elodea* and Sodium Bicarbonate. Notice that the thin tube within the rubber stopper is just barely submerged in the solution.

FIGURE 9.7

Marking the Initial Fluid Level in the Thin Stopper Tube of a Volumeter

rise when you turn on the lamp. If the fluid level falls, air is likely escaping from the volumeter, and you will need to either force the rubber stopper in more tightly or find a new stopper.

5. Wait ten minutes, and then use a plastic ruler to measure the fluid level change (in mm) in the thin stopper tube. Enter this value for *Net Photosynthesis* in Table 9.3.

Rate of Photosynthesis in *Elodea* plants

TABLE 9.3	FLUID LEVEL CHANGE (cm)
Net Photosynthesis	
Cellular Respiration	
Gross Photosynthesis	

The *Net Photosynthesis* of your plants measures the amount of oxygen that they produced while being exposed to light for ten minutes. This value does not indicate the *total* amount of oxygen produced by the plant during this period because the plant cells were also using aerobic cellular respiration to generate ATP, which requires oxygen. Net photosynthesis does not account for the oxygen used during respiration. To figure out the total amount of oxygen produced in photosynthesis by your plants (Gross Photosynthesis), you have to figure out how much oxygen they are *using* during the process. Follow the instructions below to measure the amount of oxygen used in a ten minute period by your plants.

1. Place sufficient aluminum foil around your volumeter tube so that the plants inside do not have exposure to light.

2. Place your volumeter back in the test tube rack, turn off the lamp, and mark initial fluid level in the small stopper tube using a wax pencil.

3. Allow the volumeter to sit for 10 minutes, and then make a second mark of the final fluid level (in mm). Enter the change in fluid level (in mm) for *Cellular Respiration* in Table 9.2.

Now that you have determined the amount of oxygen used by the plant cells during respiration, you can estimate the total amount of oxygen that the plants would produce in a ten minute period when exposed to direct light. To do this, simply **add** the values that you measured for net photosynthesis and gross photosynthesis. Even though your fluid level dropped in the second part of this experiment, DO NOT subtract the value from your Net Photosynthesis value!

Gross Photosynthesis = Net Photosynthesis + Cellular Respiration

Based on your results, do your plants produce excess oxygen during photosynthesis (in other words, do they produce more oxygen than they need for respiration)?

Recall that the Light Reaction, which produces oxygen, is only part of photosynthesis. If you were to measure the Calvin Cycle, what molecule would you test for in order to measure the reaction?

Compare your values for Net Photosynthesis, Cell Respiration, and Gross Photosynthesis with another lab group. List three specific variables in your experiment that could account for differences in your observations.

1. ____

2. ____

3. ____

10

CELL DIVISION

OBJECTIVES

By the end of this lab exercise, students should be able to

- describe the type of cell division that occurs in bacteria;
- list and describe the stages of the eukaryotic cell cycle;
- describe the phases of mitosis and identify them in preserved cells;
- describe how meiosis differs from mitosis in simulate meiosis using model chromosomes; and
- define **genome, chromatid, cytokinesis, mitotic spindle, fertilization, zygote, gamete, diploid,** and **haploid.**

INTRODUCTION

If you think back to your childhood days, you may recall the surprise you felt when you removed your first bandage to reveal a cut or scrape that had completely healed. You may also recall having to shop for new clothes during the years when your body was growing the most. Tissue repair and growth are two of the most important functions of cell division. When a cell divides, a *parent cell* splits into two or more *daughter cells* (see Figure 10.1). In most cases, the daughter cells produced are genetically and structurally similar (if not identical) to the parent cell, except in special cases of cell division that produce genetically unique daughter cells. In this lab, you will explore cell division by comparing the ways in which major types of cells divide.

While the process of cell division may differ greatly in distantly related cells, all dividing cells have at least two major themes in common. Before a cell can divide, it must copy its **genome.** The genome of a cell is its complete complement of DNA. Because DNA (which you will study in detail in later labs) contains the molecular instructions that allow a cell to produce the products it needs and to function in specific ways, it is important that daughter cells receive copies of the entire genome. Secondly, a cell must distribute its other components, such as organelles (in eukaryotic cells), cell membrane materials, and cytoplasm. This often requires that a parent cell grow in size and replicate many of its internal components before dividing. Beyond these simple similarities, cell division varies tremendously between major groups of cells. We will begin by working with prokaryotic cells, which exhibit a relatively simple type of division.

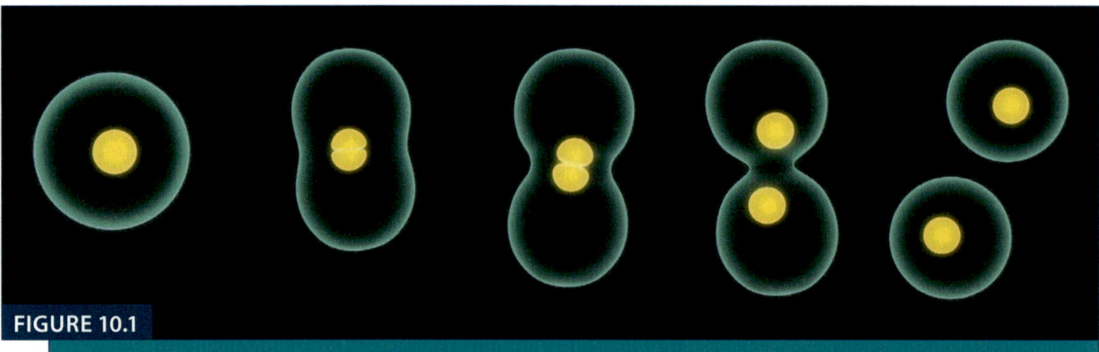

FIGURE 10.1

Cell Division. While cell division is a complex physiological process, it typically results in a parent cell dividing to form daughter cells.

PROCEDURE 10.1: CELL DIVISION IN PROKARYOTES

Recall that prokaryotic cells, like bacteria, lack a nucleus and organelles. While they do not have a nucleus, they still contain DNA, which is usually located in a central region within the cell called the nucleoid. The entire genome of prokaryotes is usually limited to one chromosome, which is often shaped like a loop. The lack of organelles, small size, and single chromosome of prokaryotes means that they tend to divide considerably more quickly than eukaryotic cells like those found in plants and animals. Prokaryotic cells divide by a process called *binary fission,* which involves replicating their looped chromosome, growing in size, and then splitting in two (see Figure 10.2). Some bacteria, like *E. Coli,* can divide in less than 20 minutes when conditions are optimal.

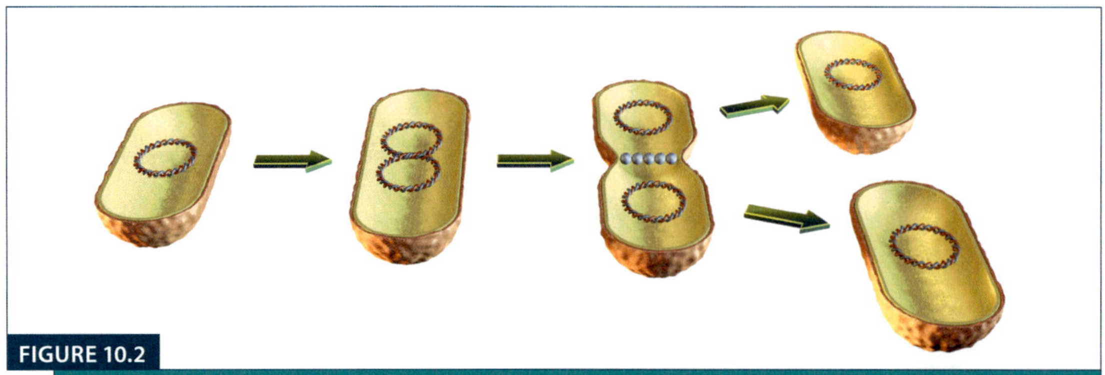

FIGURE 10.2

Binary Fission. During binary fission, bacterial cells replicate their looped chromosome and then split into two daughter cells.

We will measure the growth of bacterial colonies by binary fission over a roughly 48-hour period. By pressing your thumb onto a plastic growth plate containing a substance called agar, the bacterial cells left from your skin will have the nutrients and space they need to divide (see Figure 10.3). You will measure the rate of cell division by counting the number of bacterial colonies that form over the next two days.

While bacterial cells can divide rapidly, their division can be restricted if growth conditions are not ideal. For this procedure, you will design an experiment that tests the effects of specific chemicals on cell division. Choose a chemical from the assortment provided by your instructor that will serve as your independent variable. If you wish, you may also think of an independent variable that you think may increase the rate of bacterial division. You will be given two petri dishes containing agar. Divide the plates into quarters by drawing lines on the bottom of the dish containing the gel (see Figure 10.4). This will give you four thumbprints per petri dish, giving you eight total trials for your experiment. Use the space provided to describe how you will set up your experiment to test the effects of your independent variable on bacterial growth, including a hypothesis. Before you begin the experiment, be sure to show your instructor your protocol.

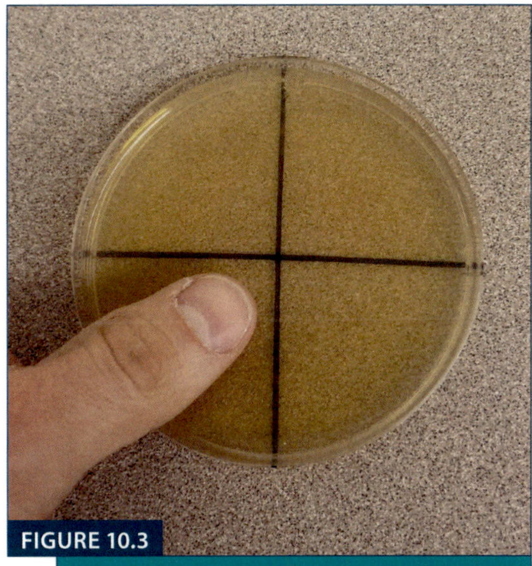

FIGURE 10.3

Transferring Bacteria. Gently place a thumb into a section on your agar plate to transfer bacteria to the plate.

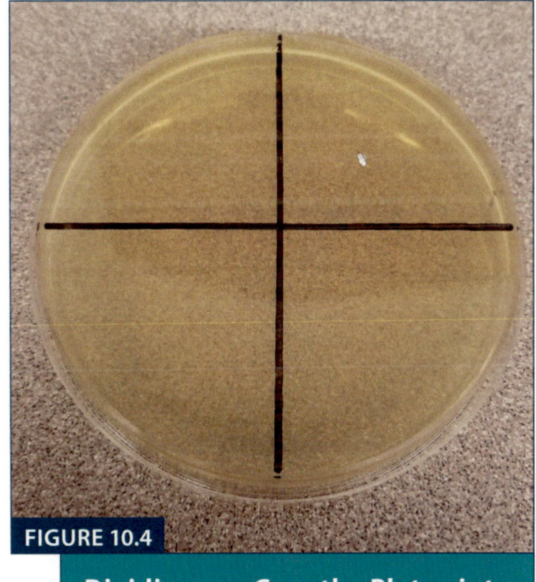

FIGURE 10.4

Dividing a Growth Plate into Quarters

EXPERIMENTAL OUTLINE: MEASURING THE EFFECTS OF __________ ON BACTERIAL GROWTH

HYPOTHESIS

When you are ready to begin, gather your needed materials and begin placing thumbprints in the different sections on the gel plate. Remember to use the same thumb for each print! Next, add your treatment to the appropriate thumbprints. When you are finished, use masking tape to seal the lid closed on the plate. Use your black marker to write your name(s) on the tape, and your instructor will incubate the bacterial cells until the next lab period.

Once your bacterial plates have incubated for roughly 48 hours, you can begin counting your bacterial colonies. Your instructor will demonstrate how to count your colonies using a colony counter (see Figure 10.5). Count the colonies in your different trial sections and record your data as a table in the space provided.

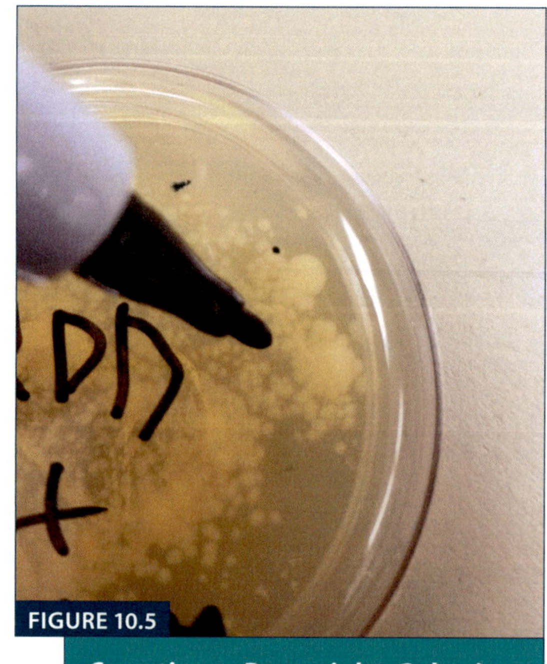

FIGURE 10.5

Counting Bacterial Colonies. Bacterial colonies can be counted using a colony counter.

Measuring the Effects of

_______________________________ **on Bacterial Growth**

Once you have counted your colonies and recorded your data, dispose of your bacterial plates in the hazardous waste containers as described by your instructor. Work with your lab partners to complete the following conclusion questions.

Did your data support your hypothesis? _______________________________

What conclusions can you draw about how your treatment affected bacterial cell division?

____

____

____

____

List three variables that you tried to control (keep constant) throughout your trials. Briefly explain how one of these variables might have impacted the growth of your bacterial cells in the different trials.

____

____

____

____

PROCEDURE 10.2: THE EUKARYOTIC CELL CYCLE AND MITOSIS

If you have ever used a washing machine to do laundry, you probably know that a typical wash cycle involves a series of steps, including rinsing, washing, and spin drying. A working washing machine will proceed through the appropriate wash steps and stop when it needs to. A broken machine may begin a new cycle after the previous cycle was complete. Eukaryotic cells, like washing machines, cycle through specific phases during their lifespan. The *eukaryotic cell cycle* (see Figure 10.6) is composed of four main stages that include specific events that occur within the cell. While prokaryotic cells also go through a type of cell cycle before dividing, it is not as complex as that seen in eukaryotes.

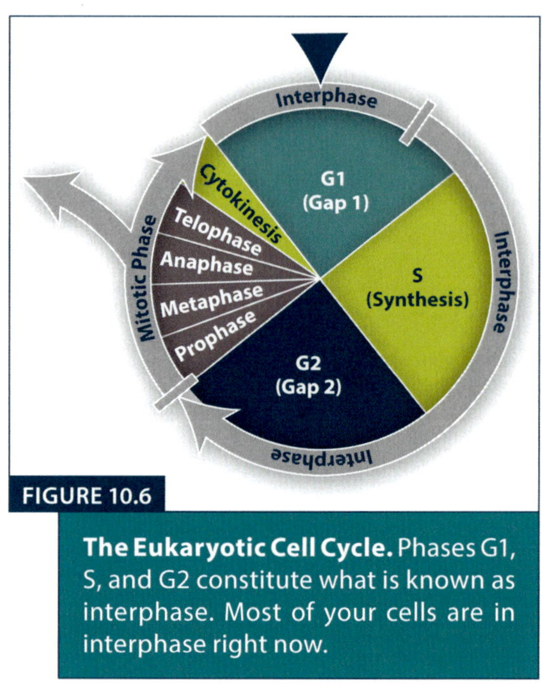

FIGURE 10.6

The Eukaryotic Cell Cycle. Phases G1, S, and G2 constitute what is known as interphase. Most of your cells are in interphase right now.

In the **G1 phase** of the cell cycle, the cell grows in size and begins synthesizing the proteins it will need for DNA replication. The "G" that appears in G1 and G2 of the cell cycle stands for "gap," as little change is noticeable in cells under the microscope when they are in these phases. A cell may "pause" during this phase and enter a state known as G0, where the cell is not actively progressing through the cell cycle. Some cells, such as muscle cells, will remain in a nondividing state indefinitely.

In the **S phase** of the cell cycle, cells replicate (copy) their genome. When a chromosome is copied, it becomes a replicated chromosome. Eventually, the cell will condense its replicated chromosomes to the point that they become visible under a powerful microscope. At this point, the chromosomes have an X shape, and they contain two sister **chromatids.** A chromatid is one half of a replicated chromosome. A replicated chromosome thus contains two sister chromatids (see Figure 10.7). The chromatids are joined together at a region called the centromere. The S in the S phase stands for synthesis, as DNA is being synthesized in the cell during replication.

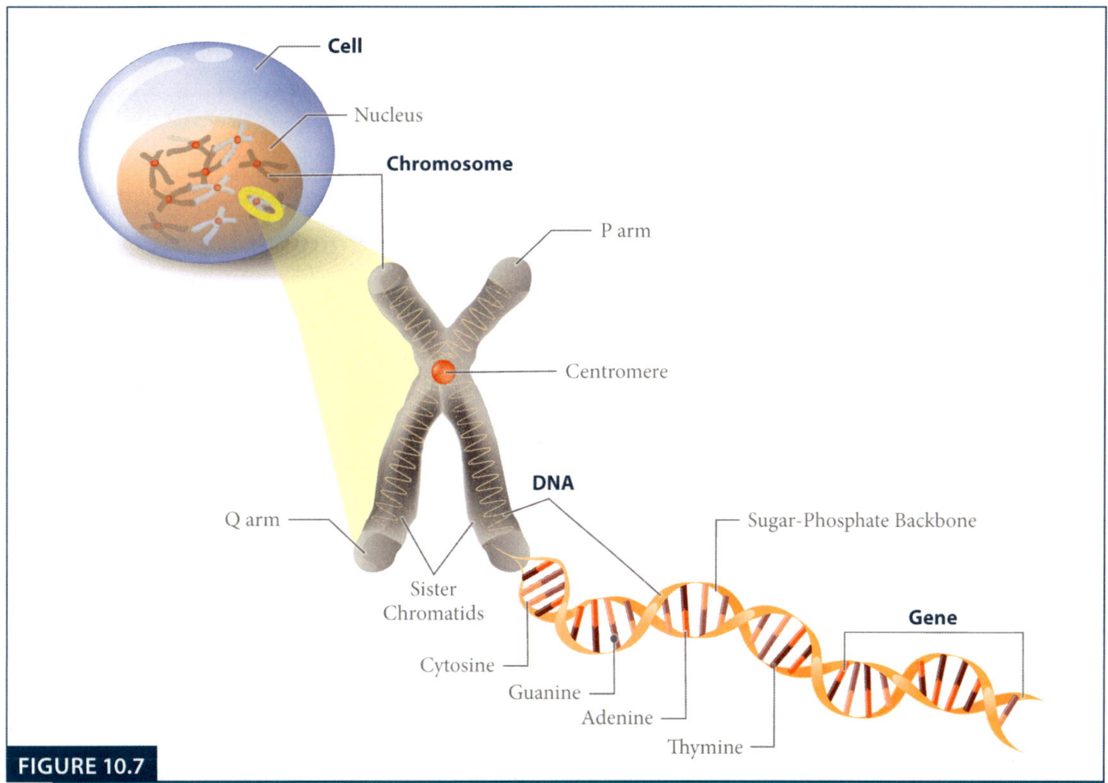

FIGURE 10.7

Chromosome Replication. After a cell copies its chromosomes, the replicated chromosomes are attached at a central point called the centromere. When a human cell completes the *S Phase*, it is still considered to have 46 chromosomes, but they are designated as replicated chromosomes, each of which contains two sister chromatids. These chromatids separate during cell division.

The **G2 phase** of the cell cycle is the second gap phase, and involves more cellular growth and protein synthesis. In this phase, the cell is replenishing energy used during the S phase, and is generating the proteins that will be needed to move chromosomes during the next phase.

The **M phase** of the cell cycle involves *mitosis,* which is the division of the cell nucleus. In many cases, the cell will split in two (a process called **cytokinesis**) after mitosis is complete. Cytokinesis is the division of cytoplasm, and occurs after the final phase of mitosis in most eukaryotes (in the case of fungi, the daughter cells formed after mitosis do not split and continue to share cytoplasm). Mitosis involves five distinct phases that you will observe under the microscope. The phases, in the order that they occur, are: prophase, prometaphase, metaphase, anaphase, and telophase (see Figure 10.8).

The cell cycle contains *checkpoints* at G1, G2, and M where a cell is able to check for errors in DNA or other abnormalities that might lead to compromised daughter cells. These checkpoints are imperative to prevent uncontrolled cell growth. Read the Thinking Critically question below before proceeding to the section on mitosis.

Thinking Critically: The Eukaryotic Cell Cycle

Not all of your cells are actively dividing. If they were, you would never be able to form functional tissues and organs. Your skin cells, for example, will begin to divide rapidly if you suffer a wound. Work with your lab partners and try to come up with an explanation as to how skin cells "know" when to divide rapidly. What do you think would happen if the cells divided uncontrollably?

You will now explore the *phases of mitosis* by viewing embryonic animal cells as they appear at each phase. We will begin by learning about the events that occur in each phase. Mitosis involves the separation of replicated chromosomes into two daughter cells (see Figure 10.8).

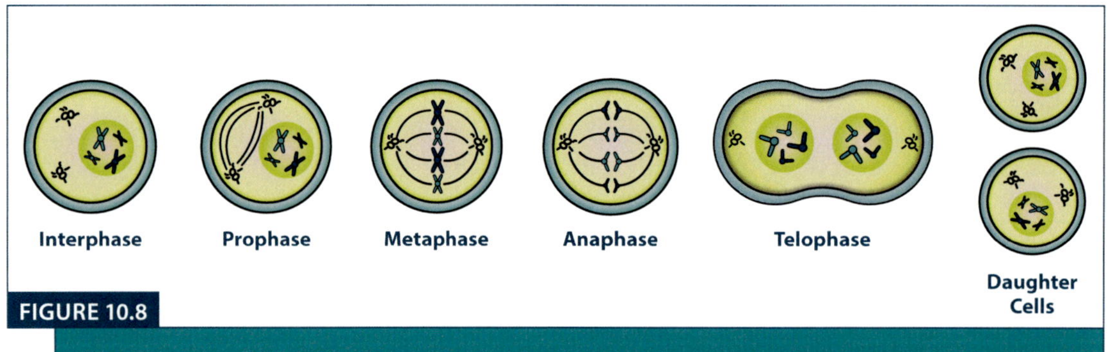

FIGURE 10.8

The Phases of Mitosis. Compare the detailed explanations of each mitosis phase with those depicted in this simplified illustration.

MITOSIS PHASES

PROPHASE

The first phase of mitosis occurs after a cell has completed G2 of the cell cycle. During this phase, the cell condenses all of its chromosomes. You can tell a cell has entered prophase because the chromosomes will be visible under the microscope. In G2 of interphase, the chromosomes in the nucleus are so long and thread-like that they appear as a jumbled mass of DNA, which shows up as one

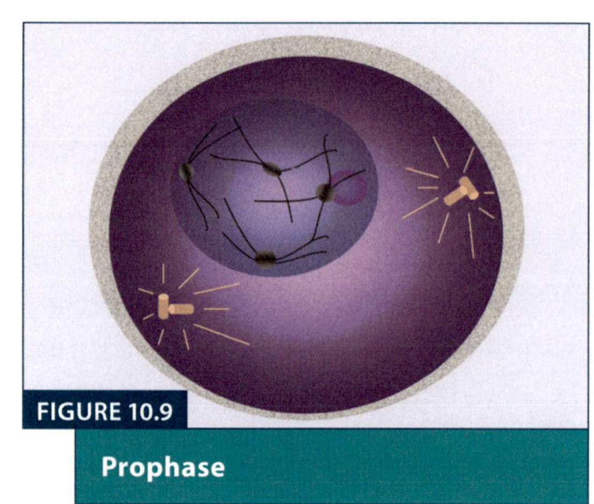

FIGURE 10.9

Prophase

solid color under the microscope. In addition to chromosomes condensing, the **mitotic spindle** forms in prophase. The mitotic spindle is an apparatus composed of microtubule proteins that move the chromosomes during mitosis. In prophase, the two organelles (called centrosomes) that form the spindle begin to move to opposite ends of the cell.

PROMETAPHASE

Prometaphase is usually considered to be the second phase of mitosis, in which the nuclear envelope breaks down completely, and spindle microtubules attach to chromosomes to begin moving them. The images in this lab activity do not show prometaphase, as it is difficult to distinguish this phase from late prophase and early metaphase under a microscope.

METAPHASE

Metaphase is the third phase of mitosis. In this phase, chromosomes are aligned along the metaphase plate, which is an imaginary line (like the earth's equator) near the middle of the cell.

ANAPHASE

In the fourth phase of mitosis, sister chromatids are separated as the spindle microtubules shorten. Once sister chromatids separate, each half of the original replicated chromosome is referred to as a chromosome.

TELOPHASE

This is the final phase of mitosis, where chromosomes arrive at opposite cell poles and begin decondensing, nuclear envelopes begin forming around the separated chromosomes, and the spindle breaks down. Many cells will split into two daughter cells in the process of **cytokinesis.**

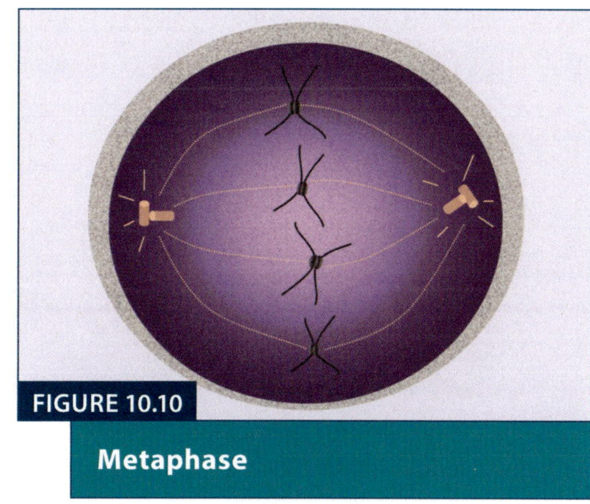

FIGURE 10.10

Metaphase

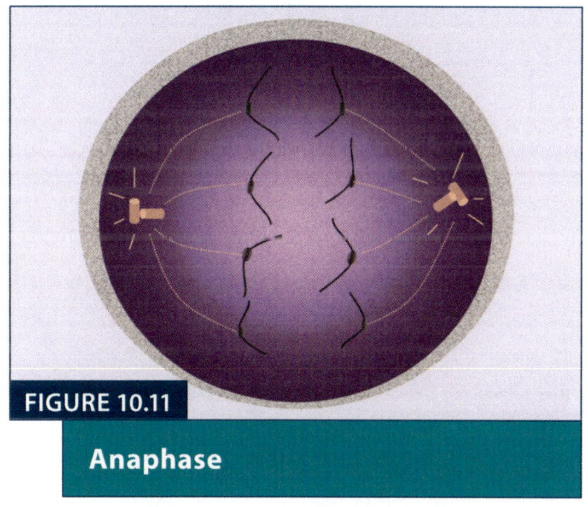

FIGURE 10.11

Anaphase

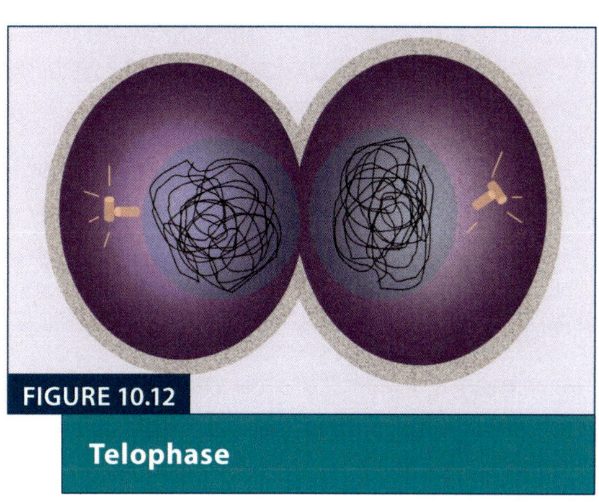

FIGURE 10.12

Telophase

Now that you are familiar with the phases of mitosis, we will now identify cells as they appear in these phases. Obtain a preserved slide of a whitefish blastula (embryo). This slide will contain a series of fish embryos which are made of around 20–30 cells that are actively dividing. Begin viewing one of the embryos on your slide at scanning power, and proceed to high power. As you observe the embryo at high power, identify how many cells you can find that are either in interphase or one of the phases of mitosis. Because it is difficult to determine if a cell is in prometaphase, we will only consider four of the mitosis phases for this exercise. Provide a brief description of the appearance of the cells that allows you to determine which phase it is in. Fill in your data in the table below.

Whitefish Cells in Different Phases of Mitosis

TABLE 10.1

PHASE	NUMBER OF CELLS COUNTED	APPEARANCE OF CELLS
Interphase (Chromosomes Uncondensed)		
Prophase		
Metaphase		
Anaphase		
Telophase		

PROCEDURE 10.3: MITOSIS IN PLANT CELLS

Recall that plant cells, like animal cells, are eukaryotic. Because they contain nuclei, plant cells can divide through mitosis in a similar manner as animal cells. While the phases of mitosis are the same in plant cells, there are some notable differences. For one, plant cells lack structures called centrioles that are found as part of the mitotic spindle in animal cells. Secondly, cytokinesis differs significantly in animal and plant cells. After telophase, animal cells begin to split by forming a *cleavage furrow.* This is a point in the parent cell where it begins to pinch in half (see Figure 10.13). When plant cells prepare for cytokinesis, a cleavage furrow does not form. Instead, plants begin to form a new cell wall, called a *cell plate,* in the middle of the parent cell after telophase. This will divide the cell in two (see Figure 10.14).

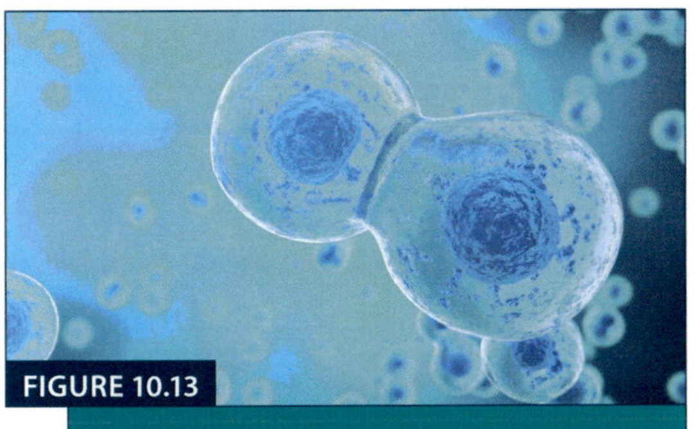

FIGURE 10.13

Cytokinesis in Animal Cells. When animal cells divide in tissue, cytokinesis occurs when a parent cell forms a cleavage furrow, which is a constricted region where the cell will pinch off into two daughter cells.

FIGURE 10.14

Cytokinesis in Plant Cells. When plant cells divide in tissue, cytokinesis occurs when a cell plate forms to separate two daughter nuclei.

Obtain a preserved slide of an onion root tip. Mitosis occurs rapidly in root tips, so you may be able to observe all of the phases in one root. As you did before, begin viewing your slide at scanning power and then proceed up to high power and answer the questions below.

What differences can you observe between the animal and plant cells pertaining to mitosis?

How might you be able to distinguish a plant cell that has just divided from a mature cell that is in interphase?

Why do you suppose that a cleavage furrow would not form in a dividing plant cell (think about how plant cells differ from animal cells)?

PROCEDURE 10.4: SIMULATING MEIOSIS AND COMPARING MITOSIS AND MEIOSIS

Like all animals, you began your life as a single cell, which had a nucleus that contained the same 23 pairs of chromosomes that most of your trillions of cells contain today. In animals, sexual reproduction involves the process of **fertilization** (see Figure 10.15), which is when a male sperm cell fuses with a female egg cell to form a **zygote,** or a fertilized egg.

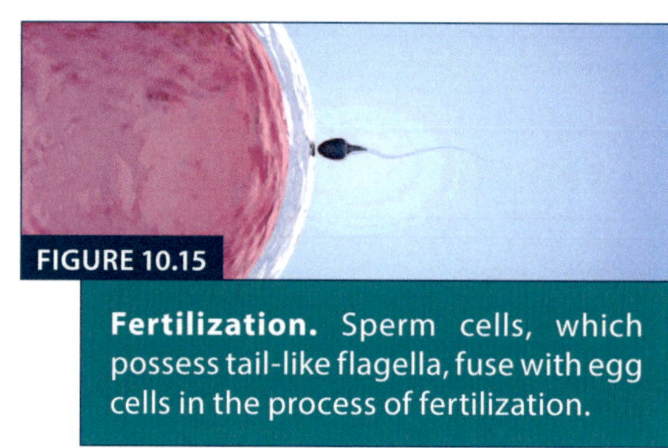

FIGURE 10.15

Fertilization. Sperm cells, which possess tail-like flagella, fuse with egg cells in the process of fertilization.

If your mother's egg cells and father's sperm cells had the normal 23 pairs of chromosomes, their fertilization would have left you with 46 pairs, or 92 chromosomes. To prevent successive generations from doubling the chromosome number of their parents, sexually reproducing organisms undergo a type of cell division called *meiosis* when generating **gametes,** or sex cells. Meiosis is a process that reduces the chromosome number of a parent cell by half. This means that human gametes contain 23 *total* chromosomes instead of 46.

In humans, meiosis occurs in *germ cells,* which, like most other animal cells, contain two copies of each chromosome. Cells are said to be **diploid** if they contain two sets of chromosomes. A diploid (also known as $2n$) human cell thus contains 23 pairs of chromosomes. Gametes are **haploid** (also known as n), which means they contain one set of chromosomes. Gametes thus do not have paired chromosomes. Females undergo a type of meiosis called *oogenesis,* which results in a diploid primary oocyte giving rise to a single haploid ovum. Males undergo spermatogenesis, which is a type of meiosis that produces four haploid sperm cells from a single diploid spermatocyte (see Figure 10.16).

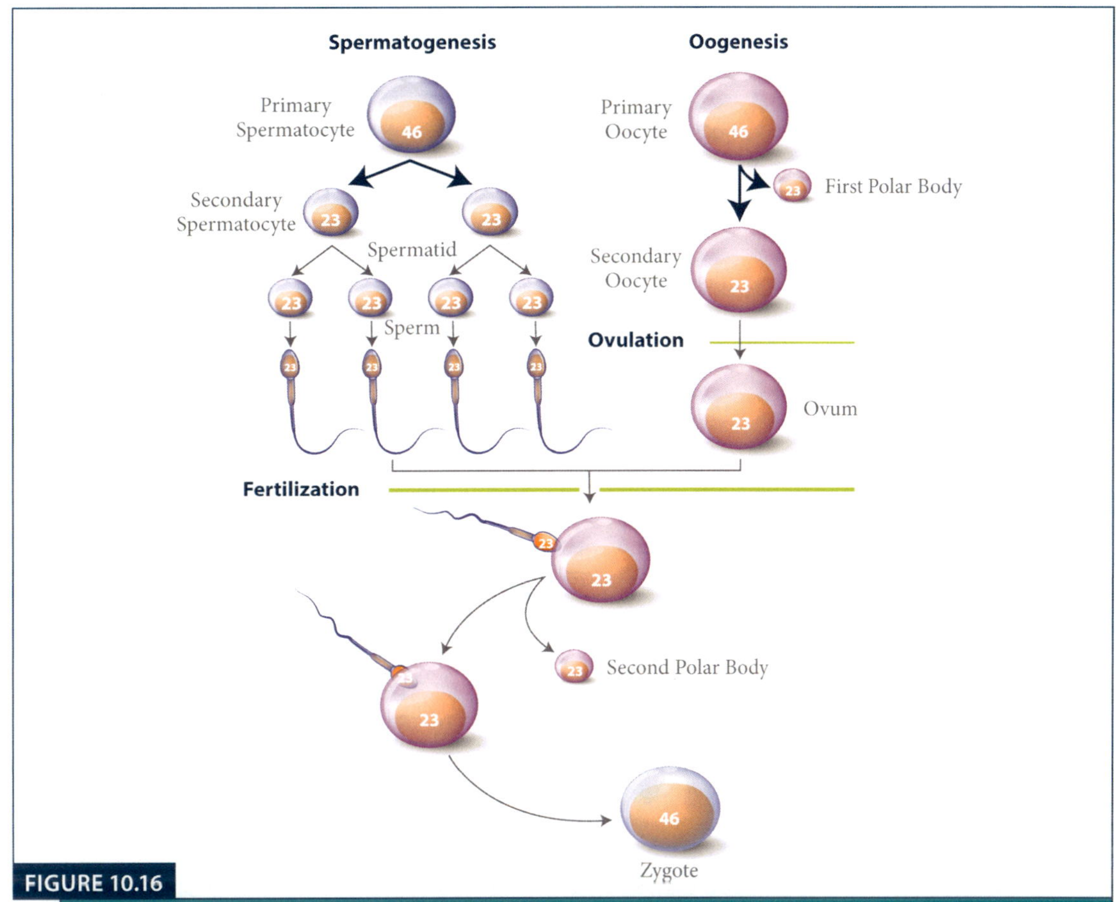

FIGURE 10.16

Spermatogenesis and Oogenesis. Spermatogenesis and Oogenesis both result in the formation of haploid daughter cells, which can eventually undergo fertilization to create a diploid (2n) zygote. While both types of meiosis produce four daughter cells, only one ovum reaches maturity. The other cells (termed polar bodies) do not reach maturity and are never fertilized.

Meiosis, unlike mitosis, involves two rounds of cell division. The phases of meiosis are thus divided into *meiosis I* and *meiosis II*. Each round of division contains separate phases, which use the same names as the phases of mitosis. Remember that your diploid cells contain 23 *pairs* of chromosomes. Each of these pairs is called a *homologous pair* because each chromosome contains the same types of genes. For each of your 23 homologous pairs of chromosomes, you received one copy from your mother and one from your father. Meiosis I involves homologous chromosomes pairing up and separating. In meiosis II, replicated chromosomes line up and separate as they did in mitosis (see Figure 10.17).

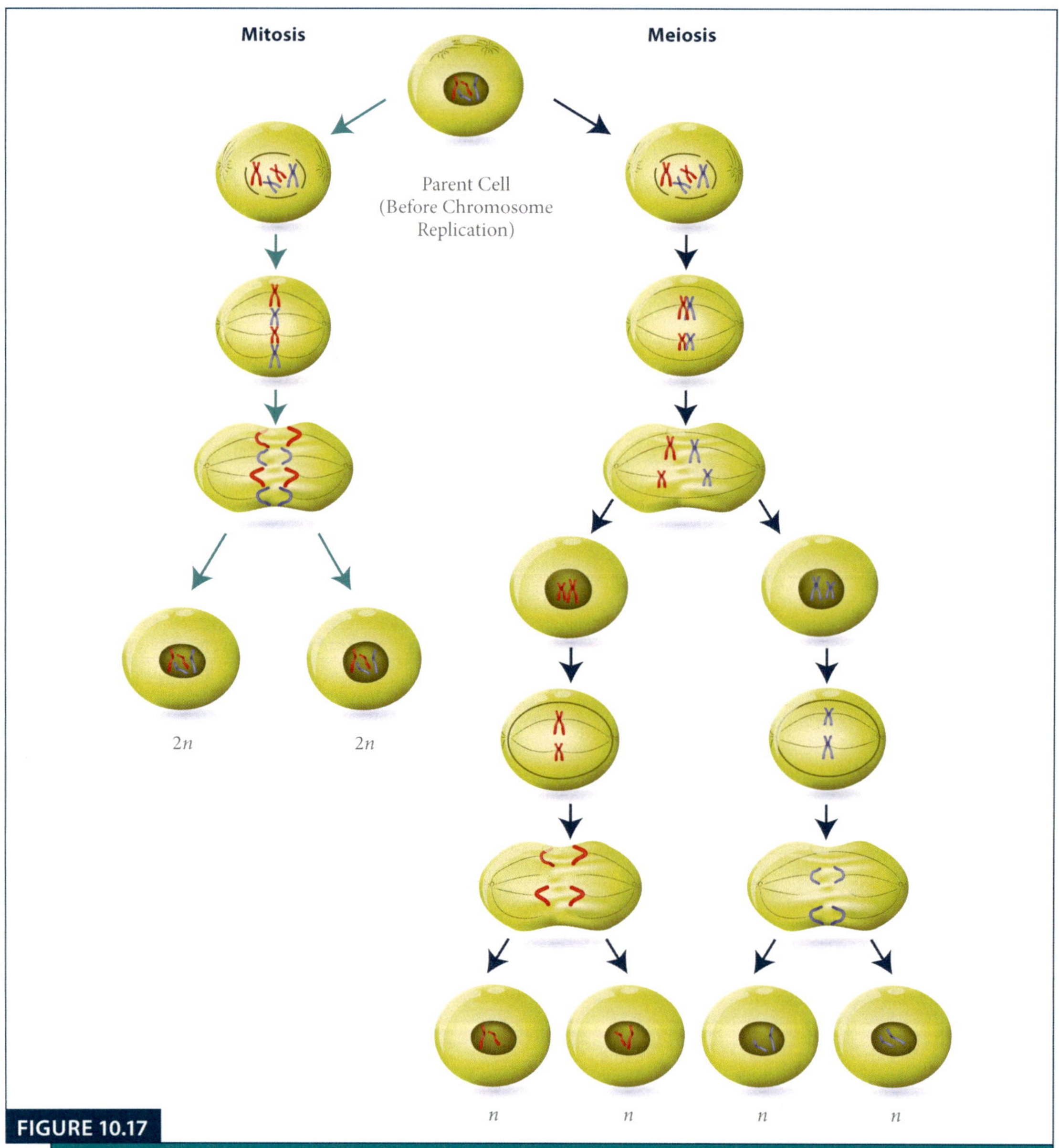

FIGURE 10.17

Meiosis. Meiosis involves two rounds of cell division, which results in four haploid (*n*) daughter cells. Both mitosis and meiosis begin with a cell that has replicated its chromosomes in the S phase of the cell cycle.

In this procedure, you will simulate the process of a germ cell undergoing meiosis using a piece of paper and chromosome models made of beads. As you proceed, refer to Figure 10.17 to help you visualize what is happening to the chromosomes at each phase.

SIMULATING MEIOSIS

1. To begin, we will assume that our parent cell has a total of four chromosomes (like the cell in Figure 10.17). The yellow and red beads will represent paternal (from father) and maternal (from mother) chromosomes, respectively. To make one pair of homologous chromosomes, attach six yellow beads to both ends of one of the tube-shaped magnetic pieces from your bin, and attach six red beads to both ends of another magnet. Keep these chromosomes separate! To make the second pair, attach three yellow beads to both ends of a third magnet, and three red beads to both ends of a fourth magnet. Place your four chromosomes on a blank sheet of paper on your table, which will represent the cell.

2. To simulate the S phase of the cell cycle, you must replicate all four chromosomes in the cell. This will mean that each chromosome will become a replicated (X-shaped) chromosome, which will be made of two sister chromatids. The sister chromatids will stick together at the magnetic portion of each chromosome, which represents the centromere.

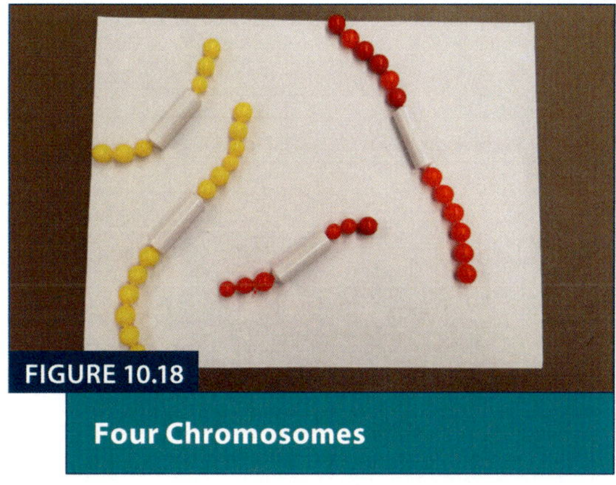

FIGURE 10.18

Four Chromosomes

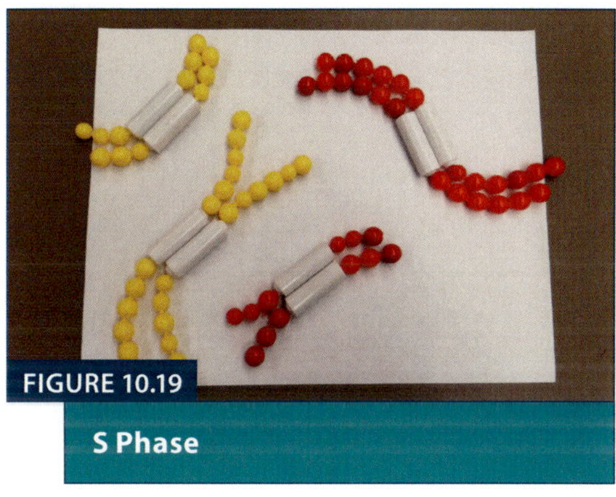

FIGURE 10.19

S Phase

3. In ***prophase I*** of meiosis, the same events that you learned about in prophase of mitosis occur in the cell, but with two important additional events. In ***prophase I*** of meiosis, homologous chromosomes pair up. Simulate this by adjoining each of your chromosomes to its homologous partner. While paired up, homologous chromosomes can undergo a process called *crossing over*. This involves maternal and paternal chromosomes swapping bits of genetic material. Simulate this by exchanging an equivalent number of beads from the yellow and red chromosomes in each homologous pair.

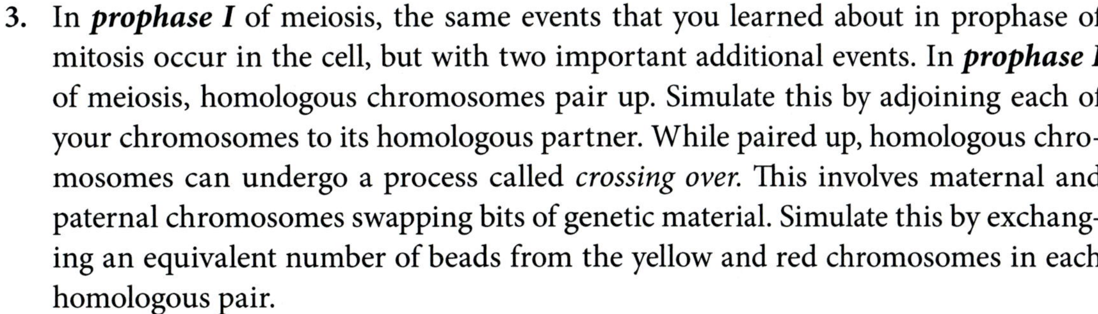

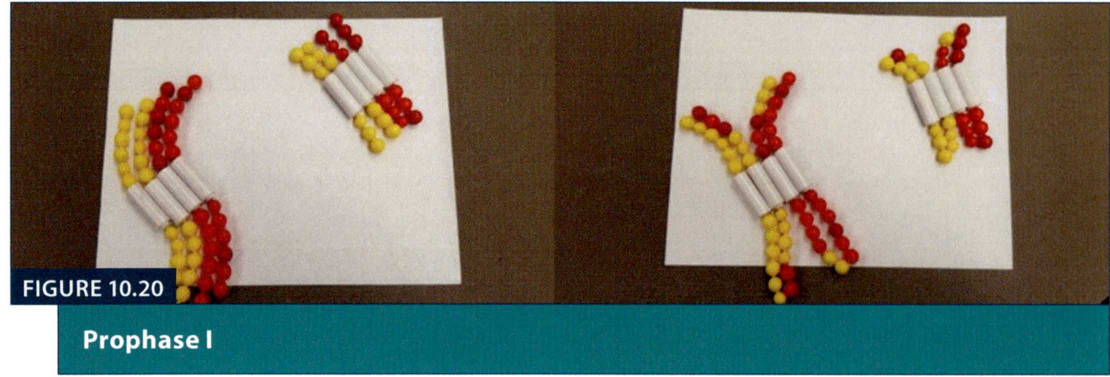

FIGURE 10.20

Prophase I

4. In *metaphase I* of meiosis, homologous pairs line up at the equator of the cell. Simulate this by lining up each pair of chromosome at the center of your piece of paper. Note that in metaphase of mitosis, individual chromosomes formed a single-file line at the cell equator.

5. In *anaphase I* of meiosis, homologous pairs of chromosomes separate. Simulate this by moving each replicated chromosome from your two homologous pairs to one side of your piece of paper (in meiosis, homologous pair separation is random, so you can move either chromosome to either side of the cell). The sister chromatids should remain attached in each chromosome!

6. In *telophase I* of meiosis, the separated chromosomes from each pair arrive at opposite ends, and the cell prepares for cytokinesis. Simulate cytokinesis by cutting your piece of paper in two, such that each half of the paper contains two of the replicated chromosomes that separated from their homologous partner. This concludes *meiosis I*.

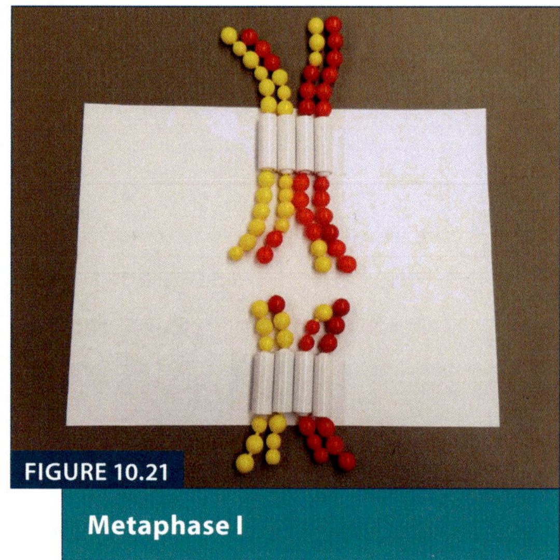

FIGURE 10.21

Metaphase I

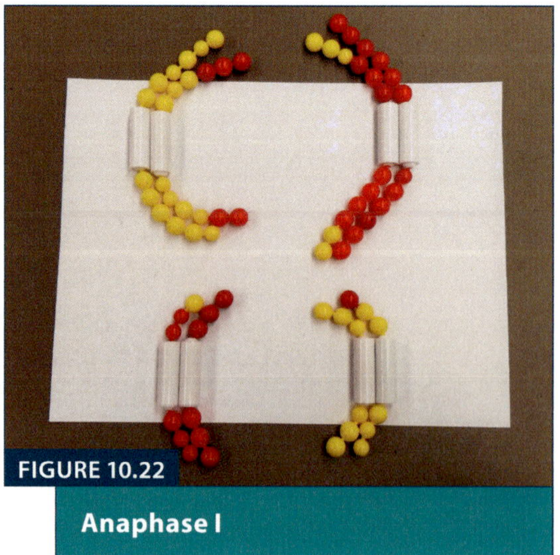

FIGURE 10.22

Anaphase I

FIGURE 10.23

Telophase I

7. We will now simulate **meiosis II** in each of the daughter cells that were just produced. The phases of meiosis are very similar to those of mitosis, beginning with **prophase II**. In this phase, the spindle apparatus forms again and attaches to chromosomes. The chromosomes will then line up in a single-file line in **metaphase II**. Simulate **metaphase II** in both of your model cells.

8. In **anaphase II,** sister chromatids separate as they did in mitosis. Simulate this in both cells by separating the chromatids at the centromere (pulling the magnets apart).

9. In **telophase II,** the separated chromatids (now called chromosomes) arrive at opposite poles of each cell, and the cells prepare for cytokinesis. Simulate cytokinesis by cutting both pieces of paper down the center a second time. You now should have four daughter cells.

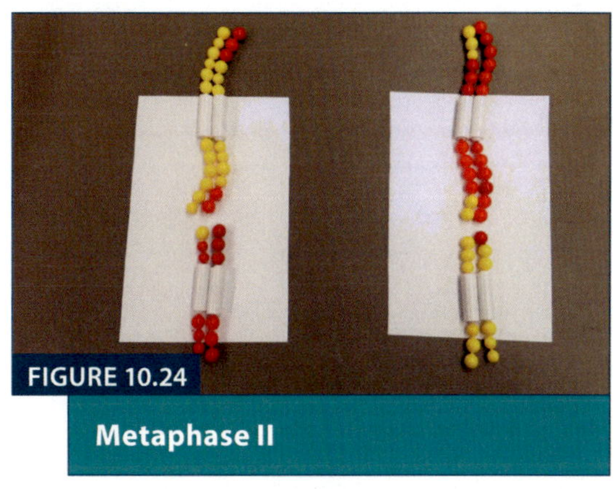

FIGURE 10.24

Metaphase II

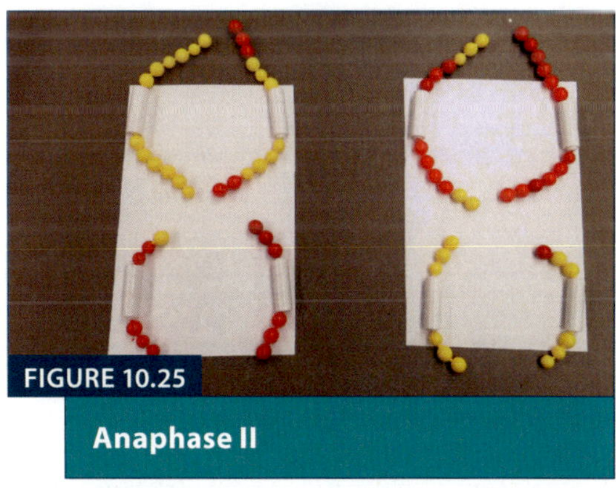

FIGURE 10.25

Anaphase II

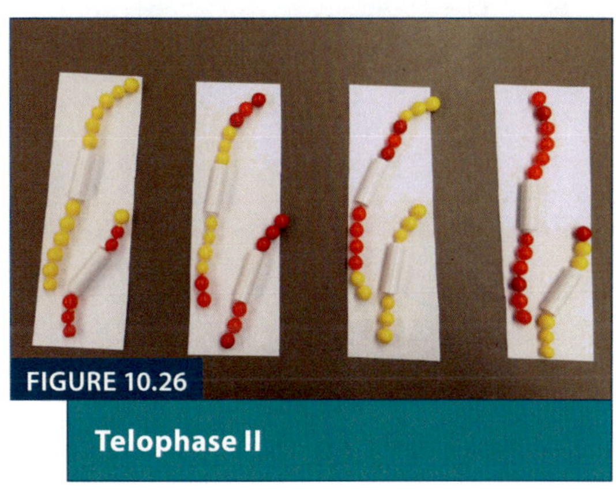

FIGURE 10.26

Telophase II

How does the chromosome number in your daughter cells compare to that of the parent cell (before it replicated its chromosomes)?

Mitosis produces two genetically identical daughter cells. Do each of your daughter cells contain the same chromosomes? Explain.

Locate a daughter cell that has a chromosome containing both red and yellow beads. Did this chromosome exist in (in identical form) in the parent cell?

Thinking Critically

Certain genetic conditions, like Down syndrome, result from an individual inheriting an extra copy of a chromosome for a given set. In the case of Down syndrome, the individual would have three copies of chromosome 21 instead of two. Other conditions, like Turner syndrome, involve an individual inheriting a single copy of a chromosome. Turner syndrome occurs in females who only inherit one copy of chromosome 23, which in females is the X chromosome. These genetic conditions result from an error that occurs during meiosis. Work with your lab partners to describe an event in meiosis I and/or meiosis II that could end up producing a daughter cell with one or three chromosomes for a set.

GENETICS

OBJECTIVES

By the end of this lab exercise, students should be able to

- understand the difference between dominant and recessive alleles;
- complete monohybrid genetic crosses for specific traits using Punnett squares;
- conduct and analyze a genetic cross in living yeast cells;
- analyze genetic pedigrees dealing with autosomal dominant, autosomal recessive, and X-linked recessive traits; and
- define **trait, gene, allele, dominant, recessive, genotype, phenotype,** and **law of segregation.**

INTRODUCTION

You probably understood from a very young age that you inherited certain characteristics from your parents, some of which you may share with your siblings. Our understanding of genetics, which is the study of inheritance, has evolved considerably since the early works of Gregor Mendel, who contributed greatly to our initial understanding of how parent organisms pass on traits to their offspring. A **trait** is any variation in the physical appearance of a heritable characteristic. As innovative as Mendel's findings were (we will study some of his work in this lab), you actually know much more about the physical makeup of a **gene** than Mendel ever did. Genes, which are units of heredity that are passed on from parents to offspring, are distinct sequences within a chromosome that code for the building blocks of proteins. Figure 11.1, which you studied in the previous lab, shows a gene as a segment of a chromosome. Remember that homologous chromosomes, of which you inherit one from each parent, carry the same (or very similar) types of genes. Genes on homologous chromosomes can become mutated over time, which can lead to multiple versions of a gene existing in a population. An **allele** is a gene variant that exists on the same location of homologous chromosomes. For example, one of the genes Mendel studied was one that contributed to the color of the seeds of pea plants. The gene exists in two allele forms: a yellow allele and a green allele. Because pea plants are diploid organisms, their cells contain two copies of the gene for seed color.

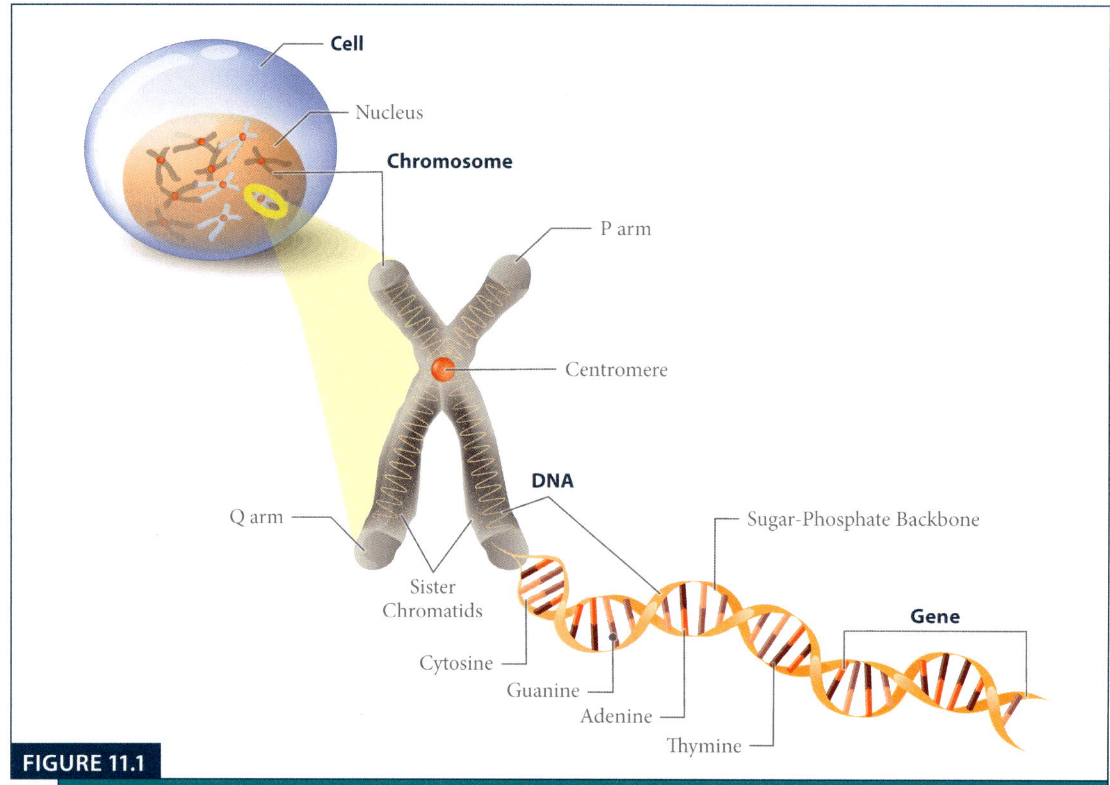

FIGURE 11.1

Genes. Genes are specific units of heredity that are found on chromosomes. Remember that chromosomes are composed of linear DNA molecules that are replicated during the S phase and condense during mitosis.

If a pea plant contains a yellow allele on one homologous chromosome and a green allele on the other, what color would its seeds be? Before the work of Mendel, many people assumed that traits in organisms resulted from a blending of those that they inherited from their parents. Mendel showed that certain alleles can *dominate* others, meaning that they mask the effects of the other type of allele. A **dominant** allele masks the effects of a recessive allele. In most cases, only one copy of a dominant version of a gene needs to be inherited for the organism to show the effects produced by that allele. In Mendel's peas, the allele for yellow seeds was dominant over the allele for green seeds. An organism must inherit *two* copies of a **recessive** allele in order for it to show up as part of their physical makeup. The **phenotype** is the observable trait expressed by an organism. Because the green allele was recessive in Mendel's peas, a plant would *only* produce green seeds if it contained the green allele on *both* homologous chromosomes.

If a single gene contributes to a trait, like in Mendel's peas, there are three possibilities for the **genotypes** pertaining to that trait. The genotype is the genetic code for a trait. A *homozygous dominant* genotype contains two dominant alleles (one on each homologous chromosome). A *homozygous recessive* genotype contains two recessive alleles. A *heterozygous* genotype contains one dominant and one recessive allele. When assigning genotypes, we usually use a specific letter to represent an allele, as letters (like alleles) can come in two versions: capital and lowercase. A dominant allele is represented by a capital letter, and a recessive allele is represented by a lowercase letter. Using the letter A, a homozygous dominant genotype would be **AA,** a homozygous recessive genotype would be **aa,** and a heterozygous genotype would be **Aa.**

Knowing about how alleles differ, explain why a homozygous dominant *or* heterozygous genotype would both produce yellow seeds (phenotype) in a pea plant.

PROCEDURE 11.1: CREATING PUNNETT SQUARES FOR MONOHYBRID CROSSES

When Mendel first crossed a parent pea plant that produced yellow seeds with a plant that produced green seeds, he may have been surprised to find that their offspring all produced yellow seeds. In this case, a plant that was homozygous dominant for seed color crossed with a plant that is homozygous recessive for seed color will indeed always produce offspring that have yellow seeds. To understand this, we have to consider the possible alleles that each parent can give to their offspring. Recall that in meiosis, chromosomes separate (or segregate) evenly into gametes, and each gamete produced receives one copy of an allele for a given chromosome set. This means that each of your gametes carries one of the two alleles for a given homologous pair of chromosomes. Mendel's **law of segregation** states that hereditary units (genes) segregate equally into gametes, so that offspring have an equal likelihood of receiving either allele from their parents. Consider, for example, a parent pea plant that is *heterozygous* for the seed color allele. We would represent this genotype with a capital and lowercase letter (i.e., Yy). When this parent plant produces its gametes, each gamete will carry *either* that dominant (Y) allele or recessive (y) allele. If a parent is *homozygous,* it can only give one possible allele type to its offspring. We can figure out the possible genotypes for offspring produced by a heterozygous (Yy) and homozygous recessive (yy) plant by creating a *Punnett square*. A Punnett square represents the possible alleles two parents can give to their offspring (see Figure 11.2).

Notice how the genotype for each parent is written either on top of the square or on the left side. The four boxes within the square represent the allele combinations that the offspring of the parents could receive. In this case, half of the offspring will produce yellow seeds as their phenotype (those that have the Yy genotype), and half will produce green seeds (those that have the yy genotype).

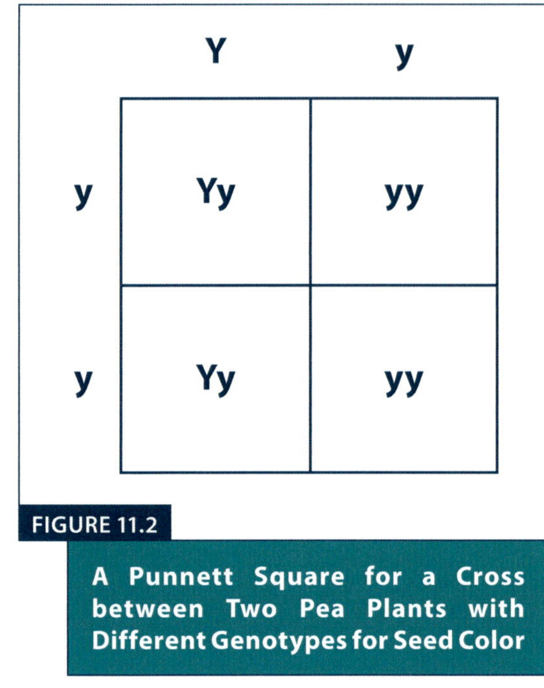

FIGURE 11.2

A Punnett Square for a Cross between Two Pea Plants with Different Genotypes for Seed Color

Practice drawing Punnett squares for the following genetic crosses:

1. Assume that, in fruit flies, red eyes (R) are dominant over white eyes (r). Draw a Punnett square that represents the genotypes possible for offspring produced by two parents who are *heterozygous* for eye color. List the possible phenotypes that the parents could produce in their offspring.

2. Draw a Punnett square that displays the genotypes and phenotypes possible for a cross between two parents who are both *homozygous recessive* for eye color.

Based on your data above, is it possible for two parent flies with red eyes to produce offspring that have white eyes? ____

What can you conclude about the eye color of offspring produced by two parent flies who are homozygous recessive for eye color?

____

____

____

Imagine you performed a cross between a parent fly that is homozygous recessive with parent of unknown genotype that has red eyes (the dominant trait). Three offspring are produced, and they all have red eyes. Would your data help you determine the genotype of the second parent (heterozygous or homozygous dominant)? Explain your reasoning below.

____

____

____

The examples we have seen so far have involved monohybrid crosses, which deal with one trait at a time. Your instructor may have you try a *dihybrid cross,* which involves a cross between two parents who have genotypes for two different traits. In a dihybrid cross, 16 possible offspring genotypes are possible.

PROCEDURE 11.2: ANALYZING PEDIGREES

Geneticists are often interested in studying the ways in which specific genetic conditions (including those that are associated with diseases) are passed on in humans. To do this, they often analyze pedigrees, which display inheritance patterns using shapes. In a typical pedigree, a white square represents a male who is unaffected by the genetic condition in question, while a white circle represents an unaffected female. Black squares and circles represent males and females who are affected by the condition

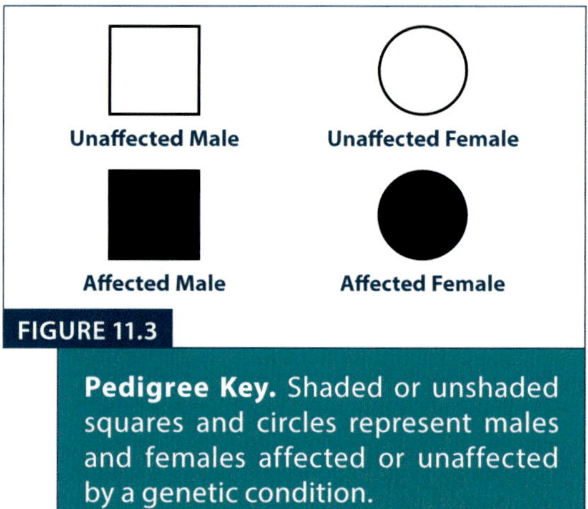

FIGURE 11.3

Pedigree Key. Shaded or unshaded squares and circles represent males and females affected or unaffected by a genetic condition.

(see Figure 11.3), respectively. A pedigree is a bit like a family tree, in that it displays the offspring of parents in a branching pattern. In this procedure, we will analyze pedigrees associated with conditions that are found on different types of alleles.

AUTOSOMAL RECESSIVE PEDIGREES

Autosomal genetic conditions in humans are those found on nonsex chromosomes (chromosomes 1–22). Because the alleles that contribute to autosomal conditions are not on sex chromosomes, males and females are equally likely to inherit them. Remember that recessive conditions will only be expressed if the individual inherits two copies of the recessive allele. In the following pedigrees, black shapes represent individuals who express (are affected by) the autosomal condition. In the case of autosomal recessive pedigrees, an affected

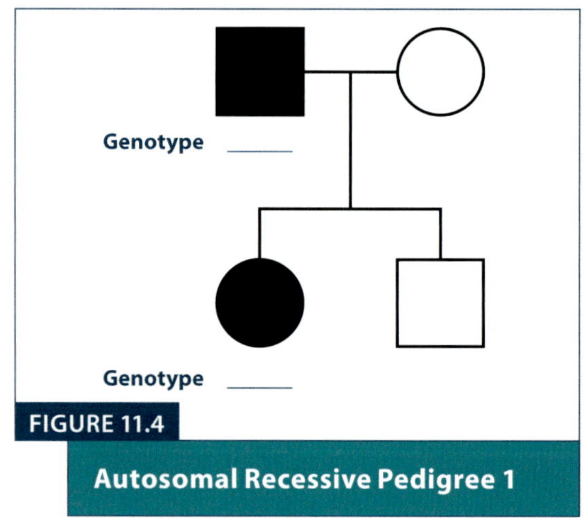

FIGURE 11.4

Autosomal Recessive Pedigree 1

individual must have inherited two recessive alleles. Using the letter A (A = dominant, a = recessive), first write the genotypes of the individuals in the pedigree who are *affected* by the condition next to the appropriate shape for Pedigree 1.

For Pedigree 1, you should have assigned homozygous recessive (aa) genotypes to the black shapes, as they are affected by the autosomal recessive condition. We know that the unaffected individuals have *at least one* dominant allele, but we are unsure of what the second allele would be (it could be dominant or recessive). When we are unsure of a genotype with a limited amount of information, we add a question mark for the second allele. In this case, "A?" would be the genotypes for the unaffected individuals.

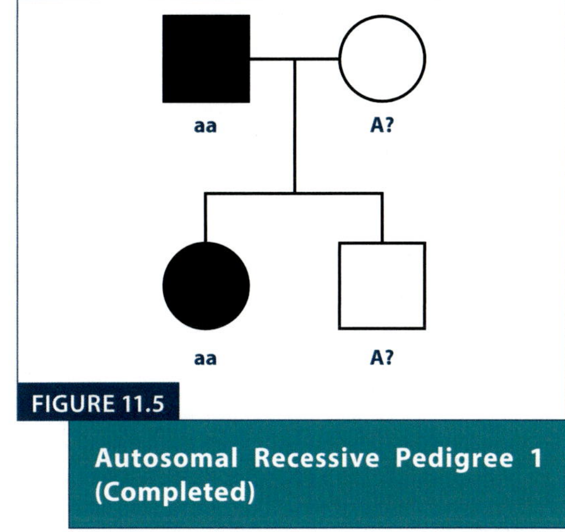

FIGURE 11.5

Autosomal Recessive Pedigree 1 (Completed)

Given that we now know some information about the alleles that the individuals in Autosomal Recessive Pedigree 1 can carry, we can assess whether or not the pedigree is possible given the genotypes of the individuals. Remember that each parent could give *either* of their two alleles to their offspring, and that each son or daughter receives one copy of their two alleles from each parent.

Is Autosomal Recessive Pedigree 1 possible?   Yes   No

If we assume that the genotype of the mother is AA for this trait, is Autosomal Recessive Pedigree 1 still possible as it is shown? Explain your response.

Complete the following autosomal recessive pedigrees by assigning genotypes to each individual and then stating whether or not the pedigree is possible.

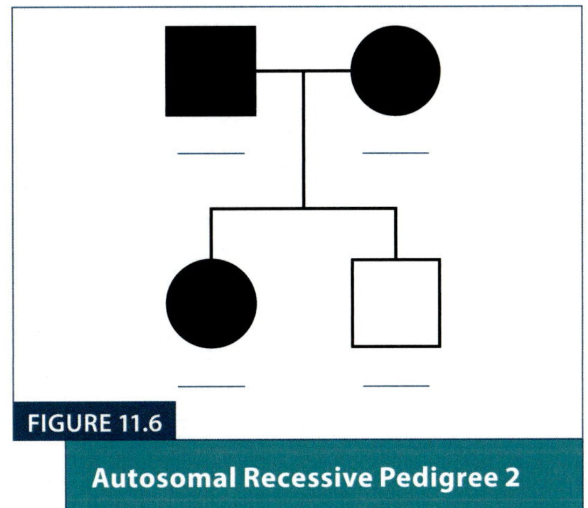

FIGURE 11.6

Autosomal Recessive Pedigree 2

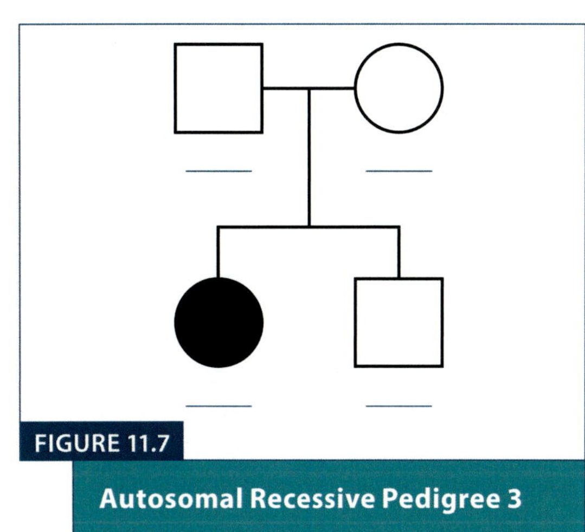

FIGURE 11.7

Autosomal Recessive Pedigree 3

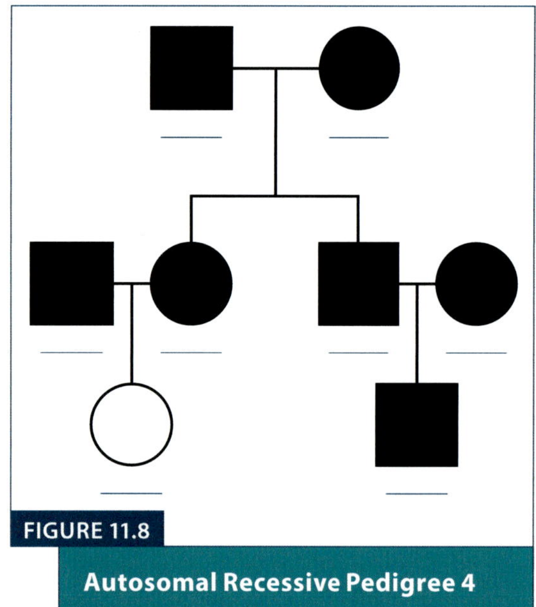

FIGURE 11.8

Autosomal Recessive Pedigree 4

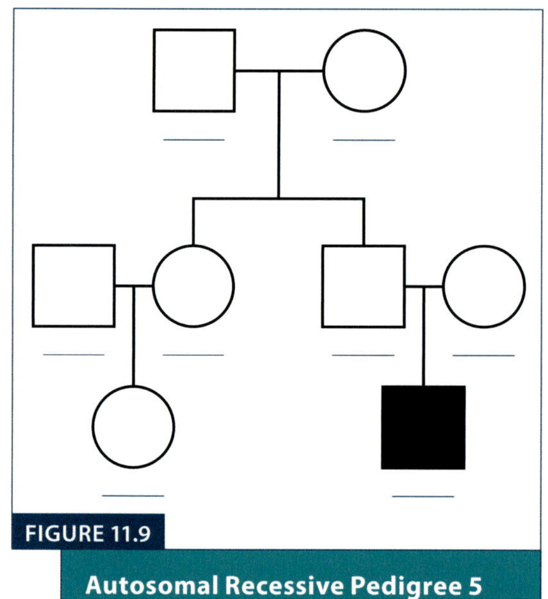

FIGURE 11.9

Autosomal Recessive Pedigree 5

Are these pedigrees possible?

Autosomal Recessive Pedigree 2   Yes   No

Autosomal Recessive Pedigree 3   Yes   No

Autosomal Recessive Pedigree 4   Yes   No

Autosomal Recessive Pedigree 5   Yes   No

Based on your analysis of Autosomal Recessive Pedigree 5, what can you conclude about the ability of autosomal recessive conditions to "skip" generations?

AUTOSOMAL DOMINANT PEDIGREES

For the next set of examples, the condition in question is on the *dominant* allele. This means that an affected individual will have *at least one dominant allele* (A). We know that in order for an individual to be *unaffected,* they most have a homozygous recessive (aa) genotype. Assign genotypes to the following autosomal dominant pedigrees, and then determine if each pedigree is possible as it is shown.

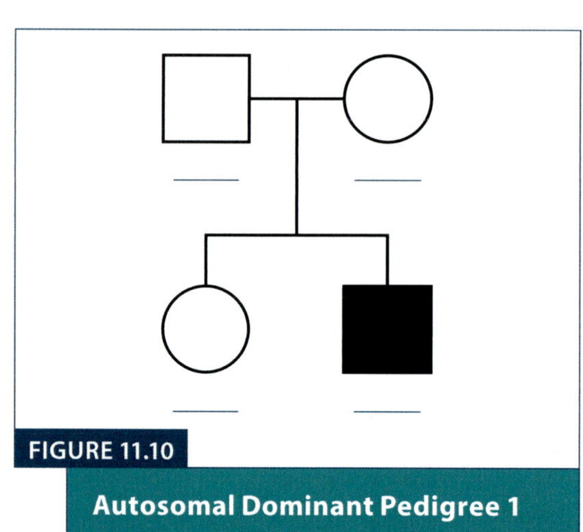

FIGURE 11.10

Autosomal Dominant Pedigree 1

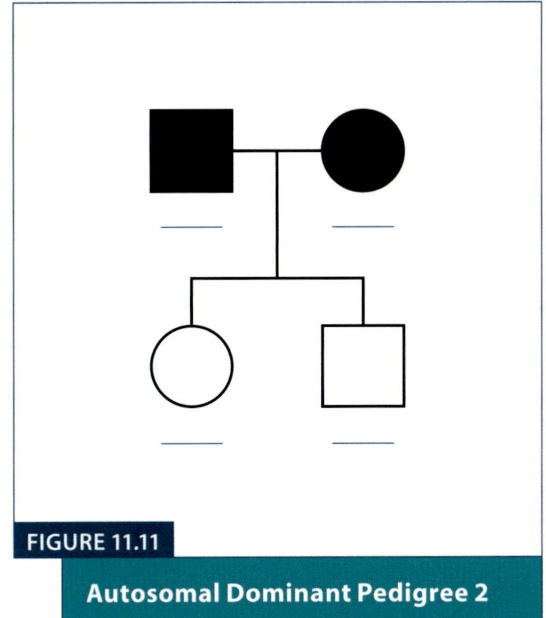

FIGURE 11.11

Autosomal Dominant Pedigree 2

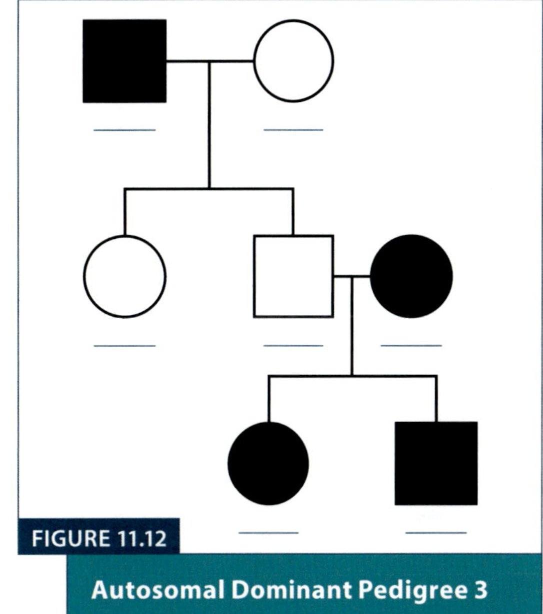

FIGURE 11.12

Autosomal Dominant Pedigree 3

Are these pedigrees possible?

Autosomal Dominant Pedigree 1   Yes   No

Autosomal Dominant Pedigree 2   Yes   No

Autosomal Dominant Pedigree 3   Yes   No

Answer the following conclusion questions regarding autosomal pedigrees.

1. What can you conclude about the offspring of two parents who are both affected by an autosomal recessive condition?

2. Is it possible for two parents who are unaffected by an autosomal dominant condition to produce offspring that are affected by the condition? ______

3. Is it possible for two parents who are unaffected by an autosomal recessive condition to produce offspring that are affected by the condition? ______

X-LINKED RECESSIVE PEDIGREES

Genetic conditions associated with alleles on the X chromosome are inherited differently in males and females. The twenty-third pair of chromosomes in humans are the sex chromosomes. Females inherit two copies of the X chromosome (one from each parent), while males inherit one copy of the X from their mother and a different chromosome, called the Y chromosome, from their father. When working with pedigrees for X-linked traits, we must take into account the genotype of each individual with regard to the sex chromosomes (XX for females and XY for males). We can still use the same letter A to represent whether the allele responsible for the genetic condition is dominant or recessive, but we will write it as a superscript above the X chromosome for each individual. Review X-Linked Recessive Pedigree 1 as an example, and then complete the pedigrees that follow as you did for the autosomal examples.

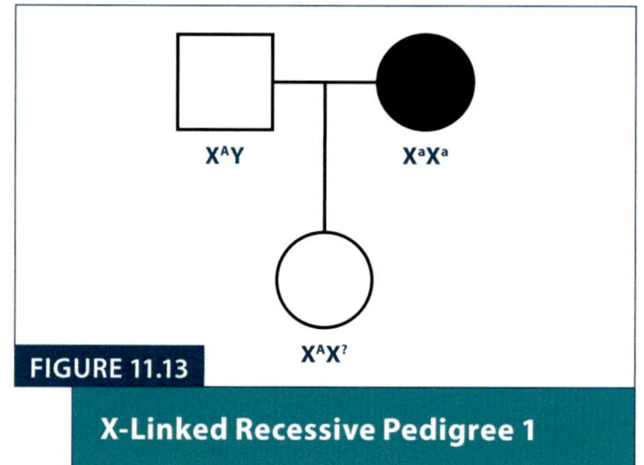

FIGURE 11.13

X-Linked Recessive Pedigree 1

In X-Linked Recessive Pedigree 1, we know that the father and daughter are unaffected, but the mother is affected. The father's only X chromosome contains the dominant allele, while the Y chromosome is unaffiliated with the condition and gets no allele above it. We know that the mother must have recessive alleles on both copies of her X chromosome, and that the daughter (unaffected) must have at least one dominant allele on one of her X chromosomes to be unaffected. This pedigree is possible, as the daughter can inherit her father's X chromosome (carrying the dominant allele) and one of her mother's X chromosomes that carry the recessive allele. As a heterozygote, the daughter would be unaffected. Keep in mind, as you complete the problems that follow, that a male only needs *one copy of the recessive allele* on his X chromosome to be affected by an X-linked recessive condition.

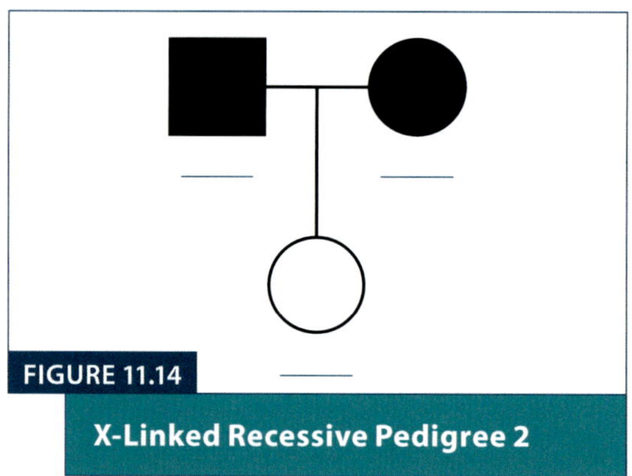

FIGURE 11.14

X-Linked Recessive Pedigree 2

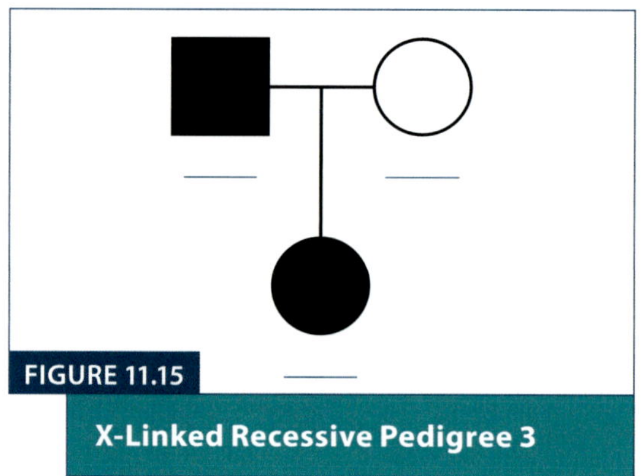

FIGURE 11.15

X-Linked Recessive Pedigree 3

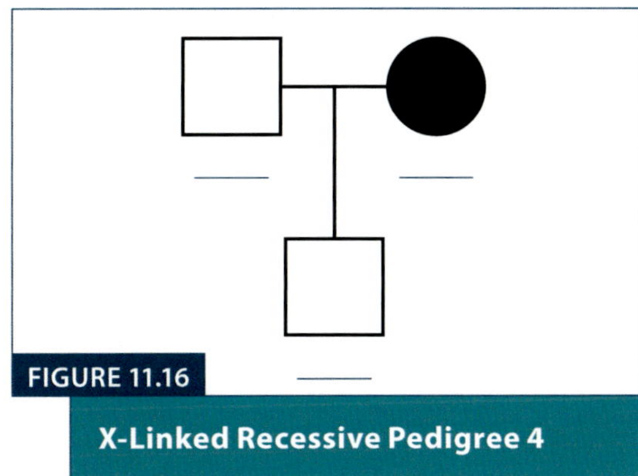

FIGURE 11.16

X-Linked Recessive Pedigree 4

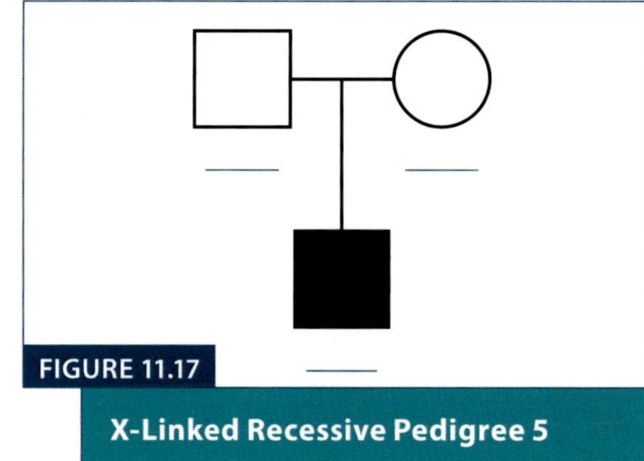

FIGURE 11.17

X-Linked Recessive Pedigree 5

Are these pedigrees possible?

X-Linked Recessive Pedigree 2   Yes   No

X-Linked Recessive Pedigree 3   Yes   No

X-Linked Recessive Pedigree 4   Yes   No

X-Linked Recessive Pedigree 4   Yes   No

Can a son inherit an X-linked trait from his father? _______________________________

If two unaffected parents produce an affected son for an X-linked condition, the mother is said to be a "carrier" for the condition. In your own words, define what you think this term would mean given the context of the X-Linked Recessive Pedigree 5 example.

PROCEDURE 11.3: CONDUCTING A MONOHYBRID CROSS IN YEAST

Yeast is a type of single-celled fungus *(Saccharomyces cerevisiae)* that humans have used to make foods and beverages for thousands of years. We have also learned a great deal about genetics from studying these organisms. Like all eukaryotes, yeast cells have a specific number of chromosomes in their nuclei. A diploid yeast cell has 16 chromosomes, but yeast, and most other fungi, alternate between diploid ($2n$) and haploid(n) phases during their life cycle. A haploid yeast cell, which results when a diploid cell goes through meiosis, has eight chromosomes. While you may be used to thinking of gametes as being either egg or sperm cells, gametes in fungi are a bit different. Fertilization in yeast occurs when haploid cells of different *mating types* fuse together and combine their chromosomes. The two mating types in our yeast samples are designated as the "a" strand and the "alpha" strand. A haploid "a" gamete will thus fuse with a haploid "alpha" gamete to produce a diploid yeast cell, much like egg and sperm gametes fuse to produce a diploid animal cell.

Yeast, like pea plants, express certain genetic traits that result from a single gene. The observable trait we will be working with will be the color of the yeast cells. In yeast, white represents the dominant phenotype for color, while red is recessive. A yeast cell will only appear red if it has two copies of the recessive allele for coloration.

Recall from the first Punnett square that you drew in Procedure 11.1 that two parents who are heterozygous for a given trait can produce offspring of different genotypes and phenotypes. We are going to perform a cross between two yeast mating types that are both *heterozygous* for the color gene. This means that, when the parent cells go through meiosis, they will produce gametes that may have the dominant or recessive allele.

DAY 1: SUBCULTURE YOUR HAPLOID YEAST STRAINS

1. Begin by obtaining a plastic YED agar plate with a circular mating grid and a container of toothpicks. Tape your mating grid to the bottom of your agar plate.

2. Lift the lid from your *alpha 1* culture plate.

3. Using the *flat* end of a toothpick from the container, gently and carefully gather a small (pea-sized) sample from the *alpha 1* culture plate and replace the lid.

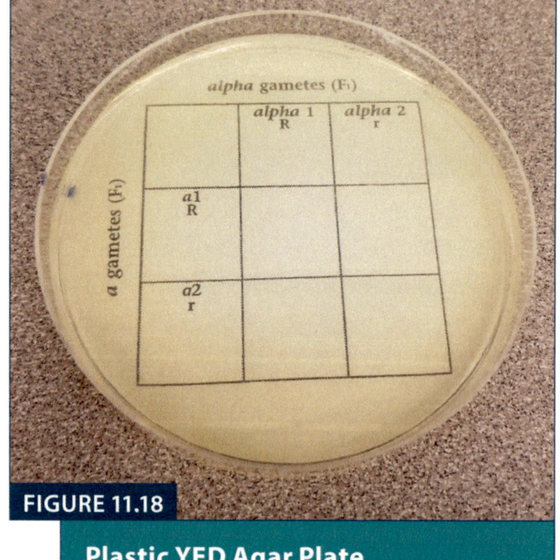

FIGURE 11.18

Plastic YED Agar Plate

4. Lift the lid of your YED agar plate and, using a circular motion, gently spot your *alpha 1* sample within the square labeled *"alpha 1/R"* (see Figure 11.18). Be very careful not to puncture the agar within your plate. Discard your toothpick and replace the lid.

5. Using a new toothpick, repeat Steps 2–4 for yeast strains *alpha 2*, *a1*, and *a2* (see Figure 11.18). Remember that each strain represents one of the two possible gametes that the heterozygous *alpha* and *a* parents can produce.

6. Write your name on the plate and incubate your plate for 24 hours. Your instructor will bring your plates to the 30°C incubator.

DAY 2: PREPARE DIPLOID MATING MIXTURES

1. Obtain your YED plate and observe the color of the haploid cells within each square.

2. Lift the lid of your YED plate and use a new toothpick to gather a small sample of the *alpha 1* strain.

3. Mark a small spot in the blank square directly below the *"alpha 1/R"* square and discard the toothpick.

4. With a new toothpick, gather a similarly-sized sample from the *"a1/R"* square on your plate. Make a small spot *next to* the spot you made in Step 3. Discard the toothpick.

5. Use another new toothpick to gently mix the spots you made in Steps 3 and 4. You should now have a mixture of the haploid gametes from the *alpha 1* and *a1* strains in one square.

6. Using a new toothpick each time, perform the same technique to mix samples of *alpha 1* and *a2, alpha 2* and *a1,* and *alpha 2* and *a2* in the three remaining squares. Discard your used toothpicks, replace the lid to your YED plate, and incubate until the next class period.

7. In the space below, draw a Punnett square representing a cross between your "alpha" and "a" yeast strains: Aa × Aa. List the ratio of phenotypes possible for this cross, remembering that white coloration (A) is dominant over red coloration (a).

DAY 3: OBSERVING RESULTS OF A MONOHYBRID CROSS IN YEAST

Obtain your YED plate and observe the results of your gamete combinations. Answer the following questions pertaining to your results.

Which gamete combination(s) produced white offspring?

White gamete combination(s) produced red offspring?

Does the ratio of the offspring phenotypes in your plate match those that you predicted in your Punnett square?

If you combined haploid cells from your red yeast culture with those of another group's red yeast culture, what percentage of the offspring produced would you expect to be red?

12

DNA STRUCTURE, FUNCTION AND ISOLATION

OBJECTIVES

By the end of this lab exercise, students should be able to

- describe the structure of DNA and the types and components of DNA nucleotides;
- explain how DNA polymerase enzymes function in DNA replication;
- describe how DNA sequences can code for specific amino acids;
- differentiate between DNA and RNA;
- conduct an experiment to extract DNA from living cells; and
- define **nucleotide, purine, pyrimidine, helicase, topoisomerase, antiparallel, transcription, translation, codon,** and **mutation.**

INTRODUCTION

We have studied different aspects of DNA in various labs so far, including its location in eukaryotes within cell nuclei, its replication and segregation during cell division, and the manner in which genetic traits are inherited in the previous lab. We have now come to the point where we can take a closer look at the structure of DNA, as well as the ways in which it functions in cells.

We now know that DNA (deoxyribonucleic acid) is the universal unit of heredity in organisms, in that it provides the genetic instructions that cells use to build protein products. To understand how it is that DNA is able to function in this way, we need to refer back to the concept of macromolecules, which you may recall exist as polymers (chains of smaller molecules called monomers). The monomer of a DNA molecule is called a **nucleotide.** DNA nucleotides contain three important components: a five carbon sugar called *deoxyribose,* a phosphate group, and one of four different types of bases. The four DNA bases are adenine, guanine, thymine, and cytosine. DNA nucleotides can polymerize to form a chain when a bond forms between the sugar of one nucleotide and phosphate group of another. Unlike the other polymers we have studied, DNA is *double stranded.* This means that one chain of DNA nucleotides can form a bond with another, which produces a helical (spiral-shaped) ladder called a double helix (see Figure 12.1).

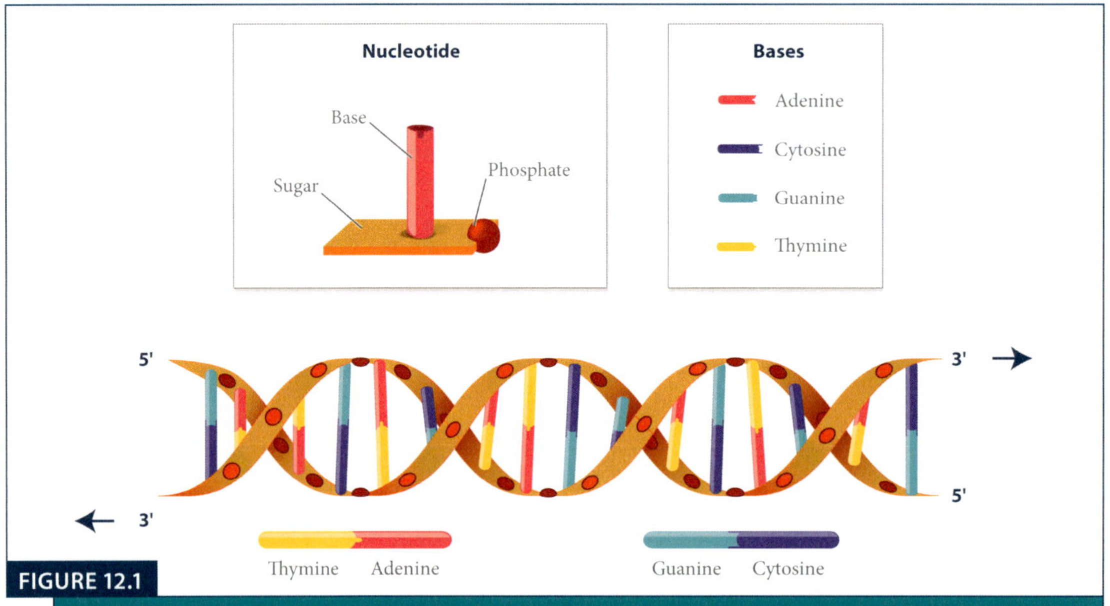

FIGURE 12.1

DNA Structure. DNA nucleotides come in four varieties based on the base they contain. Nucleotides chain together to make a double-stranded helix.

If the shape of DNA were to be thought of as a twisted ladder, the sides of the ladder would be made of alternating sugar/phosphate groups. This is referred to as the sugar phosphate backbone of DNA. The rungs of the ladder are formed by two adjacent bases, which form hydrogen bonds to hold the two DNA strands together. Notice in Figure 12.1 that adenine is always bonded to thymine, while guanine is always bonded to cytosine. This is due to a concept called *complementary base pairing.* In a DNA molecule, adenine always bonds to thymine, while cytosine always bonds to guanine. The structure of the complementary bases gives the DNA helix a uniform width: adenine and guanine are **purine** bases, and they contain two carbon rings in their chemical structure. Cytosine and thymine are **pyrimidine** bases, which only contain one carbon ring (see Figure 12.2). The width of the helix is thus always defined by a purine base bonded to a pyrimidine base, and never two purines (which would make the helix thicker) or two pyrimidines (which would make the helix thinner).

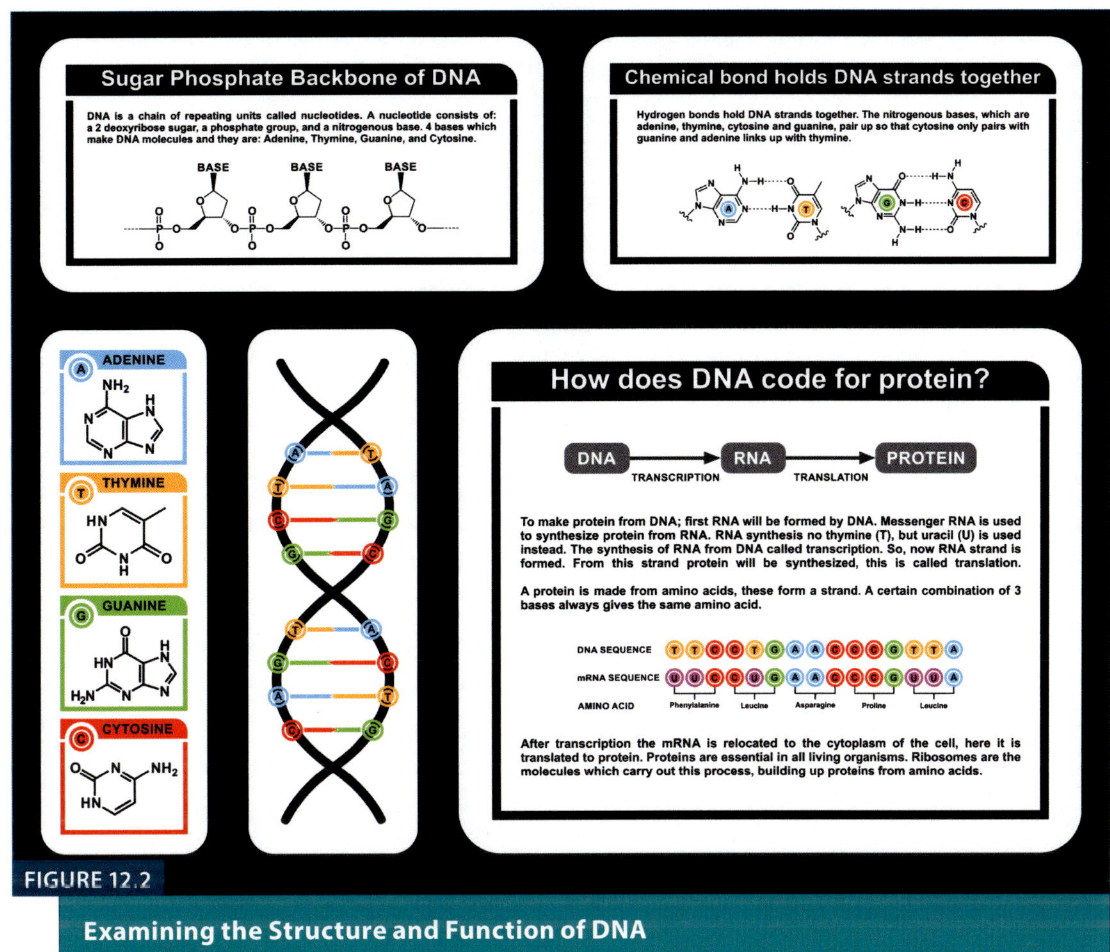

Examining the Structure and Function of DNA

PROCEDURE 12.1: OBSERVING DNA STRUCTURE AND SIMULATING DNA REPLICATION

Observe the DNA model at your table, and notice the helical shape formed by the two adjacent strands. The model contains base pairs that are abbreviated with the letters A, T, C, and G.

How many purine bases do you count in the model? _______________________

How many pyrimidine bases do you count? _______________________

Is the width of each bonded base pair the same throughout the DNA molecule? _______

Beginning with the top of your model, follow one of the two strands to the bottom, and write the sequence of bases that occur from top to bottom below.

Base Sequence of Model DNA Strand 1: ________________________________

Without looking at the model, write the sequence of the second strand based on what you now know about complementary base pairing.

Base Sequence of Model DNA Strand 2: ________________________________

While the model does not indicate the differences in complementary base pair bonds, notice in Figure 12.2 that adenine and thymine are bonded by two hydrogen bonds, while cytosine and guanine are bonded by three hydrogen bonds. It requires more energy to separate the bond between cytosine and guanine than that between adenine and thymine. The ability for hydrogen bonds between DNA bases to break is imperative when cells enter the S phase of the cell cycle, as they must replicate all of their DNA. In order for a DNA molecule to be replicated, the bonds between the two adjacent strands must be separated. Two adjacent DNA strands can be thought of as a zipper, as two sides that make up a zipper on a coat can be bonded together or separated. In order for DNA to begin to be copied, the two strands must be unwound ("unzipped") at a point called the *origin of replication*. In eukaryotes, there are several origins of replication for each chromosome. The origins are identifiable because they contain a specific sequence of bases.

Based on what you know about the strength of the complementary base bonds in DNA, which base pair would you expect to occur more frequently in an origin of replication (where the DNA must initially be opened up)?

____

____

Once two strands are unwound at the origin, the enzyme **helicase** is responsible for unwinding and opening the two strands of the helix in a specific direction. Helicase can be thought of as the zipper that unzips a coat (see Figure 12.3). Another enzyme, called **topoisomerase,** associates with the DNA molecule ahead of helicase in order to relieve tension that is applied to the molecule by helicase. As helicase unwinds the DNA, the region ahead of the enzyme can become supercoiled, and this is prevented by the activity of topoisomerase (see Figure 12.3).

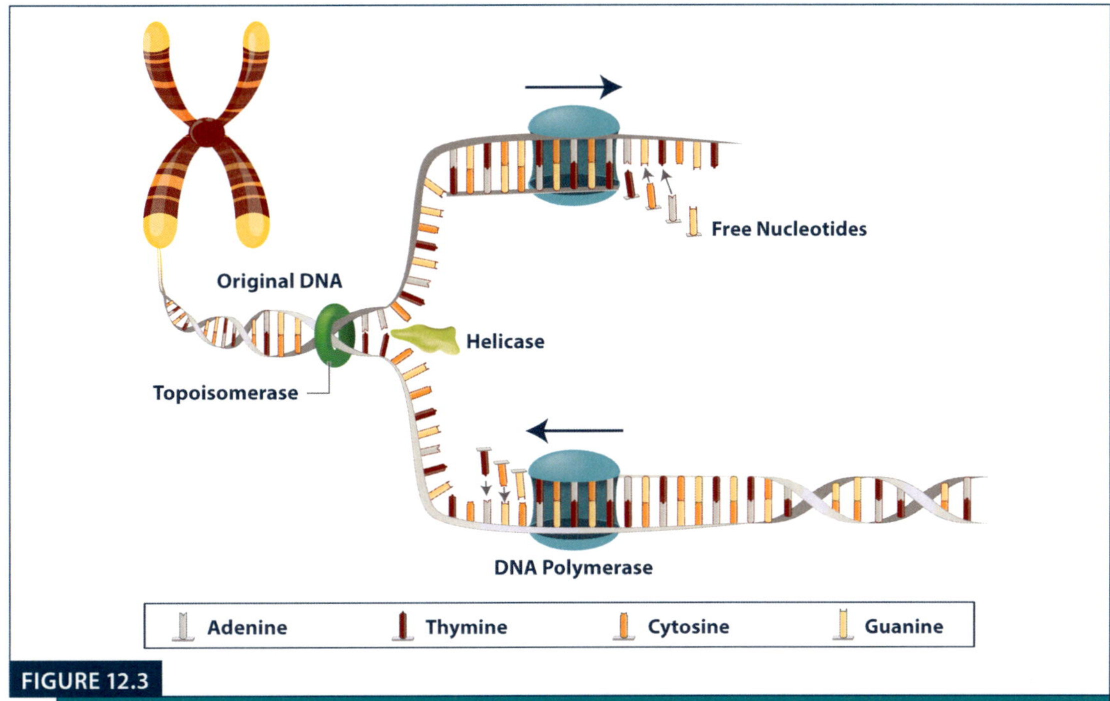

FIGURE 12.3

DNA Replication. Helicase unwinds DNA strands by breaking apart hydrogen bonds between base pairs, while topoisomerase prevents supercoiling of the DNA molecule ahead of helicase.

After the double helix is unwound, two new complementary strands are added to the two original strands by *DNA polymerase* enzymes. Once the entire chromosome has been unwound and replicated, the two new DNA molecules will each contain one strand from the old molecule and one new strand. This means that DNA replication is *semiconservative,* meaning part of the old molecule is conserved in each new DNA molecule.

Notice in Figure 12.3 that the two DNA polymerase enzymes are working in opposite directions. We now know that a DNA molecule contains two adjacent strands held together by hydrogen bonds, and that the two strands are equally spaced apart (parallel). Each strand has two specific ends, referred to as the 5' end and the 3' end (see Figure 12.1). The 5' end of one strand on a DNA molecule always faces the 3' end of the other, meaning the strands line up in a "head-to-tail" fashion. DNA is said to be antiparallel, meaning the strands are equally spaced but line up in opposite directions. DNA polymerase enzymes are limited in how they are able to synthesize new DNA; they can *only* build a new complementary strand of DNA in the 5' to 3' direction. Consider the DNA sequence in Figure 12.4.

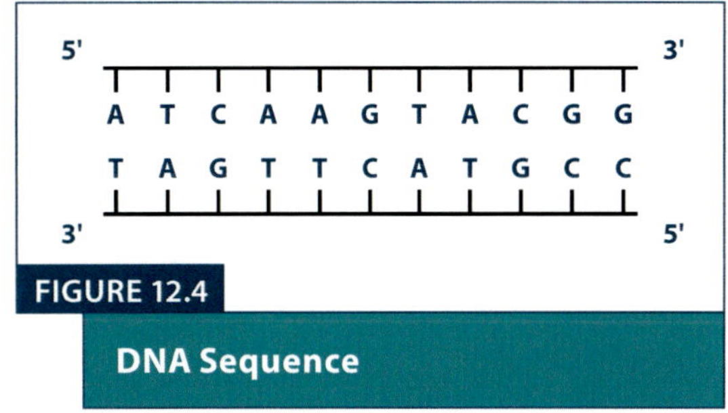

FIGURE 12.4

DNA Sequence

Imagine that helicase is unwinding the DNA from the left to the right of the page. This will separate the two strands so that two new strands can be built. However, DNA polymerase must move in different directions on the two strands being copied. On one strand, polymerase can follow behind helicase as it is building new DNA in the 5' to 3' direction, but on the opposite strand, polymerase must work backwards (away from helicase) to build new DNA. The strand that is being built in the direction of helicase is the *leading strand,* which is made faster than the *lagging strand,* which is built away from helicase. In the space below, draw the unwound DNA strands with a colored pencil, and then draw the two new complementary sequences that would be built from the old with a different colored pencil. Indicate the 5' and 3' ends of each strand that you draw, remembering that the two new DNA molecules must both be antiparallel.

In your drawing, indicate which of the two new strands that you drew would be the *leading strand,* and which would be the *lagging strand* (remember that DNA polymerase can only build a new DNA strand beginning with its 5' end and ending with its 3' end).

PROCEDURE 12.2: TRANSCRIPTION AND TRANSLATION

The genetic code of all cells is written in the form of DNA, and the language of DNA is a sequence of nucleotide bases contained within a chromosome. We know that our genetic code is what our cells use as a blueprint for making protein products, but we have yet to explore the manner in which genes lead to proteins. The *Central Dogma of Biology* states that information in cells is processed from DNA to RNA to proteins. RNA, which stands for ribonucleic acid, is another type of nucleic acid that is similar to DNA. DNA differs from RNA in a few key ways, including the overall structure of the polymers. DNA molecules are double-stranded, while RNA molecules are single-stranded. RNA also contains the sugar ribose, whereas DNA contains deoxyribose. RNA and DNA share the bases adenine, cytosine, and guanine, but RNA nucleotides can contain the base uracil (U) instead of thymine (see Figure 12.5).

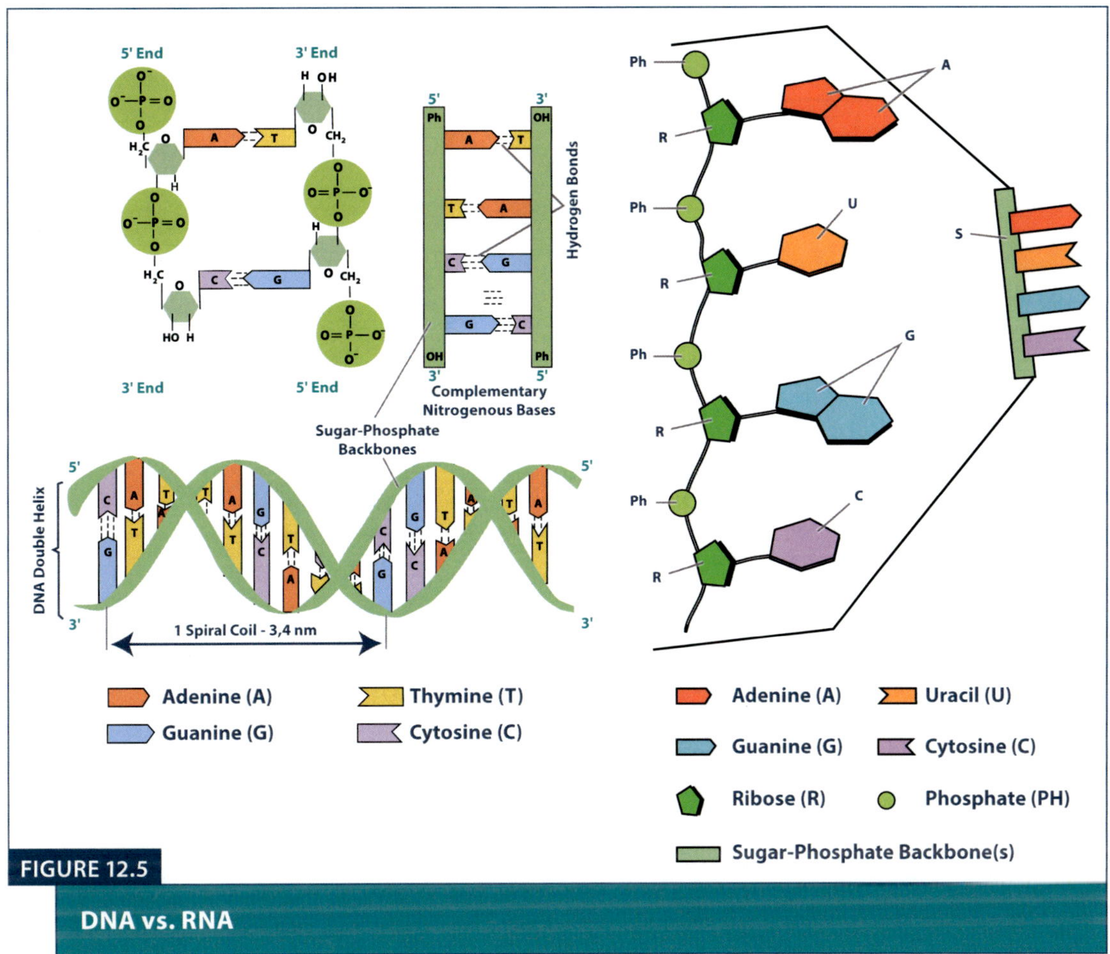

FIGURE 12.5

DNA vs. RNA

Recall that ribosomes manufacture proteins, and that proteins are chains of amino acids. While a ribosome builds a specific protein based off of the coding in DNA, it cannot "read" the DNA message itself. Ribosomes, which are themselves made of a type of RNA, produce proteins by reading sequences of RNA. A DNA code must be converted into an RNA code before a protein can be made. The process of **transcription** involves a sequence of DNA (a gene) being converted into a sequence of RNA. An enzyme called RNA polymerase binds to the DNA molecule, unwinds the DNA strands, and begins synthesizing a complementary sequence of RNA. RNA bases can form complementary bonds with DNA similar to how two DNA strands can bond together, but with the base U (uracil) replacing T (thymine). If a DNA strand contained the sequence ATCG, the complementary RNA sequence would be UACG.

Like DNA polymerase, RNA polymerase must build a nucleotide chain from the 5' to 3' direction. When RNA polymerase locates the beginning of a gene, it begins to build a sequence of RNA based off of one of the two DNA strands (called the template strand). Consider the DNA strand sequence in Figure 12.6.

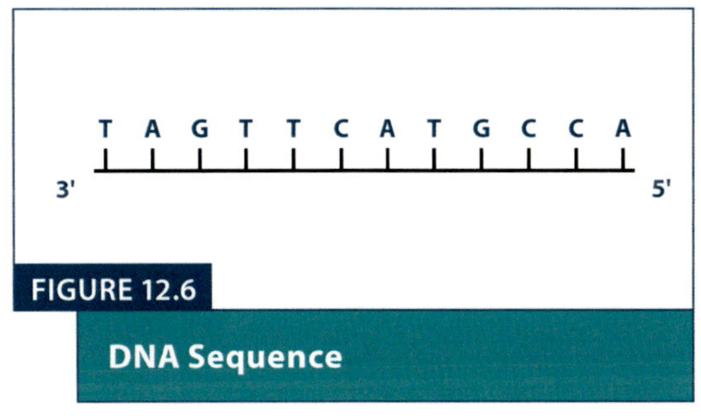

FIGURE 12.6

DNA Sequence

If this strand were being transcribed by RNA polymerase, write the complementary RNA strand that would be synthesized.

RNA Strand Sequence Transcribed:

The strand of RNA produced in transcription is called the *mRNA transcript.* The "m" in mRNA stands for messenger, as this type of RNA carries a message to the ribosome. After transcription, the mRNA transcript travels to the ribosome. In the process of **translation,** the ribosome builds a chain of amino acids based off of the sequence in the mRNA transcript (see Figure 12.7). The resulting chain of amino acids will be modified in various ways to become a functional, three-dimensional protein.

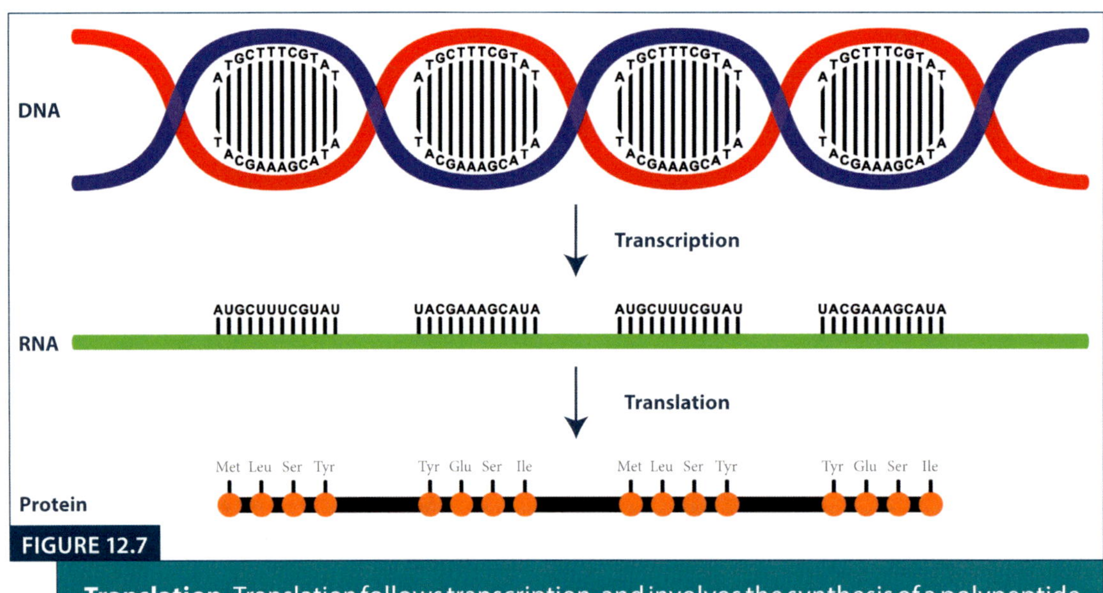

FIGURE 12.7

Translation. Translation follows transcription, and involves the synthesis of a polypeptide chain from an mRNA transcript.

The mRNA is read three bases at a time in sequences called *codons.* Your transcript above contains four codons (12 bases total). To figure out the chain of amino acids that a ribosome would produce based off of your mRNA transcript, we can refer to a chart (Figure 12.8) that shows the amino acids produced by different codons. Using Figure 12.8 as a reference, write the sequence of amino acids that would be produced from the four codons found in your mRNA transcript.

FIGURE 12.8

Codon Chart

Second Letter

First Letter	U	C	A	G	Third Letter
U	UUU UUC } Phe UUA UUG } Leu	UCU UCC UCA UCG } Ser	UAU UAC } Tyr UAA Stop UAG Stop	UGU UGC } Cys UGA Stop UGG Trp	U C A G
C	CUU CUC CUA CUG } Leu	CCU CCC CCA CCG } Pro	CAU CAC } His CAA CAG } Gln	CGU CGC CGA CGG } Arg	U C A G
A	AUU AUC AUA } Ile AUG Met	ACU ACC ACA ACG } Thr	AAU AAC } Asn AAA AAG } Lys	AGU AGC } Ser AGA AGG } Arg	U C A G
G	GUU GUC GUA GUG } Val	GCU GCC GCA GCG } Ala	GAU GAC } Asp GAA GAG } Glu	GGU GGC GGA GGG } Gly	U C A G

Amino acid sequence: ___________ ___________ ___________ ___________

Thinking Critically: Regulating Transcription

All of your somatic cells that have a nucleus contain the same DNA, but different cells transcribe different genes in your DNA depending on their needs. A *transcription factor* is a molecule that regulates the rate at which a gene is transcribed, or if it is transcribed at all. Think of a time in your life when you began to produce gene products that you were not producing when you were younger. What chemicals/products do you think your body began producing that could have influenced the transcription of these genes?

Keep in mind that the amino acid sequence you produced in the previous example is ultimately determined by the original sequence found on the DNA strand that you transcribed. If the DNA sequence were changed, you might end up with a different amino acid produced during translation, which might cause the resulting protein to lose its function (or take on a different function in the cell). A **mutation** is a permanent change in a DNA base sequence. A *point mutation* occurs when a single base is swapped for another. For example, imagine that the original DNA sequence from the previous example had the third base "G" replaced with a "T."

What codon would be produced from this three-letter DNA sequence after transcription? ____

What amino acid would be produced from this codon after translation? _______________

In this case, you should have produced the same amino acid as the original DNA sequence from the first example. Many point mutations are *silent,* meaning they result in the same amino acid as the original sequence would have produced (there are 20 amino acids, but 64 possible codons).

A *frameshift mutation* occurs when a base is inserted or deleted from a DNA sequence, which causes the entire downstream "sentence" of codons to be changed. Refer to the DNA sequence from the first example and insert an extra "A" base before the first "T" base. Write the RNA sequence that would be produced from this mutated DNA sequence, and then translate the first four codons in the mRNA into an amino acid chain.

mRNA transcript produced from mutated DNA sequence:

Amino acid chain produced from mutated DNA sequence:

How many of the amino acids in this chain are different from the sequence that you produced from the original (not mutated) DNA? ____

PROCEDURE 12.3: ISOLATING AND EXTRACTING DNA

In this procedure, you will extract DNA from strawberry cells. Wild strawberries normally have two copies of seven chromosomes (14 total in the diploid state). Through selective breeding, many strawberries sold in grocery stores have eight copies of each of the seven chromosomes (56 total). Using the following protocol, you will be able to isolate the DNA from store-bought strawberry cells, which you can transfer to a microcentrifuge tube (filled with ethanol) and take home (see Figure 12.9). In order to access the DNA, we must first break apart the cell membranes, which will require the use of a detergent. Once isolated using the right chemicals, the DNA from the plant cells will collect and float at the top of your solution, at which point you can extract it from the tube using a sterile loop.

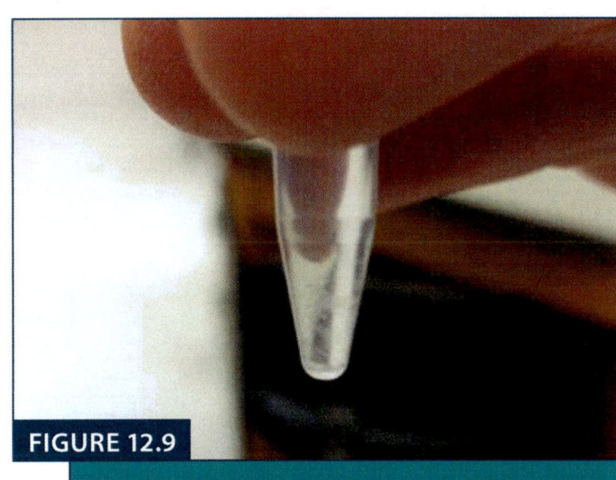

FIGURE 12.9

DNA Extract. Your DNA extract will appear as a cloudy layer of bubbles, which you can transfer into a microcentrifuge tube.

DNA EXTRACTION STEPS

1. Add 10 mL of distilled water to the plastic bag containing your strawberry sample.

2. Mash the strawberry until all of the solid lumps are gone.

3. Place a plastic funnel with two layers of cheese cloth or gauze into the large test tube.

4. Cut a small hole in the corner of your bag and slowly drain the fluid into your funnel. (It is important that you do not get any material into the tube that hasn't been filtered!)

5. Measure 5 mL of filtered juice into the small (capped) test tube.

6. Add 1 mL of 8% NaCl into the small test tube.

7. Add 5 drops of dish soap to the small test tube.

8. Gently invert the test tube three times. (Try not to allow bubbles to form!)

9. Slowly add 5 mL of cold ethanol to the small test tube.

10. Carefully place small test tube in the test tube rack and allow DNA to separate to the top of the tube.

11. If you would like to extract and keep your DNA sample, use the sterile plastic loop to scrape the DNA out of the container and transfer it into a plastic vial that is half filled with ethanol.

Thinking Critically: Extracting DNA

While we can be confident that most of the DNA that you extracted came from the strawberry chromosomes, we cannot be certain that DNA from other organisms might have been extracted as well. What other DNA might your sample contain, and how do you think it could have gotten into the strawberry cells?

13

GEL ELECTROPHORESIS AND DNA FINGERPRINTING

OBJECTIVES

By the end of this lab exercise, students should be able to

- describe how restriction enzymes are used in DNA fingerprinting;
- create an agarose gel plate for use in electrophoresis;
- explain how gel electrophoresis can be used to compare DNA samples;
- conduct an experiment that simulates a crime scene investigation using gel electrophoresis; and
- define **restriction enzyme, palindrome,** and **RFLP.**

INTRODUCTION

Just as every individual has a unique fingerprint, they also have unique sequences of DNA nucleotides that make up their chromosomes. Sometimes, as in the case with forensic analysis, investigators need to be able to compare the DNA of different individuals. Unlike fingerprints, which can be seen by the naked eye and easily compared, chromosomes and their component DNA nucleotides are not easily visualized, even under powerful microscopes. Thus, even though DNA provides a genetic fingerprint for an individual, comparing DNA samples requires sophisticated technology and techniques. In this lab, we will learn about a process called agarose gel electrophoresis, which can be used to compare DNA samples from different individuals. This technique involves loading DNA samples into a gel plate, which are then separated using electrical charge (see Figure 13.1).

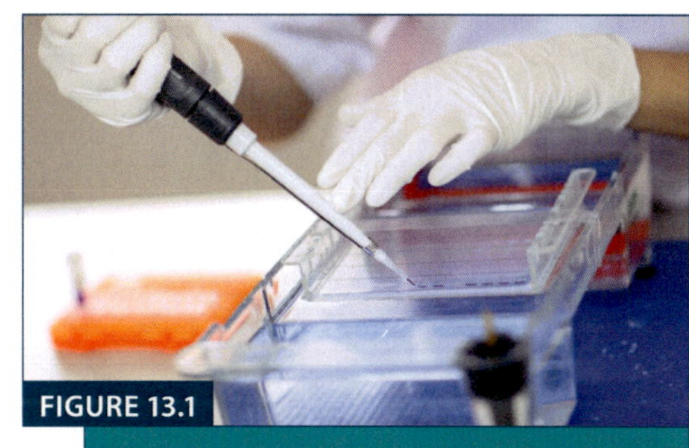

FIGURE 13.1

Loading DNA Samples. DNA samples are loaded into a gel plate before they are filtered using electrophoresis.

Agarose gel electrophoresis is a technique that uses an electrical field to separate DNA fragments based on both their electrical charge and relative size. In order to separate DNA fragments, we can use a substance called agarose, which acts as a "molecular sieve" to filter DNA fragments of different sizes (see Figure 13.2).

Because agarose is porous in its gel form, small fragments of DNA can travel more quickly through the material than large ones, which travel more slowly. We will be

FIGURE 13.2

An Agarose Gel Plate with Separated DNA Fragments

using agarose to filter DNA fragments that have been treated with *restriction enzymes.* Restriction enzymes are able to make "cuts" in DNA molecules at specific sequences. Specifically, they catalyze the cleavage of the phosphodiester bonds that form between the sugar and phosphate molecules of adjacent nucleotides in a DNA strand. A restriction enzyme cuts DNA by finding a recognition site called a **palindrome.** In a DNA molecule, a palindrome is a sequence that reads the same from 5' to 3' on opposite strands. The sequence GAATTC is an example of a palindrome that acts as a recognition site for a specific restriction enzyme. Consider the following DNA sequence:

5' ACAAGAATTCATTAGA 3'

3' TGTTCTTAAGTAATCT 5'

Notice that GAATTC is read the same from 5' to 3' on opposite strands. When a restriction enzyme finds a specific palindrome, it cuts the DNA at that location, creating two fragments of DNA. We must remember that everyone has unique DNA base sequences within their chromosomes. If we treated a specific chromosome from two different individuals with a restriction enzyme that cut DNA molecules at every GAATTC sequence, we would expect a different number of fragments to be produced between the two based on how many times GAATTC showed up on their chromosomes. Consider the following two DNA sequences from two individuals:

Individual 1

5' ACAGGTACGAATTCAGGACTCAGTACCGTA 3'

3' TGTCCATGCTTAAGTCCTGAGTCATGGCAT 5'

Individual 2

5' CGAGAATTCAGGTACAGTAGAATTCAGATA 3'

3' GCTCTTAAGTCCATGTCATCTTAAGTCTAT 5'

If a restriction enzyme cut these two DNA samples at the GAATTC site, how many DNA fragments would be left for individual 1? _______________________

How many fragments would be produced for individual 2? _______________________

If you then filtered the samples through a sieve, which individual's DNA fragments would you expect to travel further/faster through the gel? _______________________

The size of DNA fragments thus depends on the spacing between two palindrome sequences. A DNA fingerprint can thus be generated between two individuals by using restriction enzymes to produce **RFLPs,** or Restriction Fragment Length Polymorphisms. RFLPs are the result of variations in length of a given segment of DNA between two recognition sites (palindromes). In this lab, we will be comparing synthetic DNA that has already been treated by two different restriction enzymes to produce DNA fragments. These samples are in vials labeled A–F, and will be described later. It will be your job to make the agarose gel, load the DNA samples into the gel, run the electrophoresis experiment, and then analyze the results. Our simulation contains samples of crime scene DNA and DNA samples from two different suspects (no real DNA is used in this lab experiment).

PROCEDURE 13.1: PREPARING THE AGAROSE GEL

For our experiment, we will be preparing a 0.9% agarose gel plate. We will make 100 mL of agarose gel, but this will be more than you will need to make the gel plate (your instructor will help you to determine exactly how much of the agarose to pour when making your plate).

1. Begin by filling your graduated cylinder with 100 mL of 1× buffer solution from the carboy in the back of the room.

2. Weigh exactly 0.9 g of agarose powder using a small, plastic weigh boat and one of the electronic balances at the sides of the room.

3. Transfer your buffer solution into your small beaker, and then add the agarose powder. Stir the solution using a stirring rod, and place the beaker in a microwave for **one minute.**

4. Remove the beaker *with gloves* and stir the solution. Put it back in the microwave for 30–40 seconds. When you remove it (with gloves) the second time, your solution should be clear. If it is not, stir it with your stirring rod, and heat it for another 20 seconds if needed.

5. Allow your beaker to cool at your table for 10–15 minutes, or until your instructor gives you permission to proceed to the next step.

6. Obtain a rubber dam device with a casting tray (see Figure 13.3). Place the casting tray in the center of the dam device and tighten the device to lock the tray in place. Make sure that the tray is flush with the bottom of the device. Fill the tray with water from the sink to make sure that the rubber dams are fluid-tight. If you do not notice any fluid leaking from the tray, you may dump the water down the sink drain and wipe the tray dry with a paper towel.

7. Place your comb into the slots on your casting tray (your instructor will demonstrate where to place the comb), and begin pouring your cooled agarose gel into the casting tray. Your instructor will help you determine how much agarose you should pour into your tray.

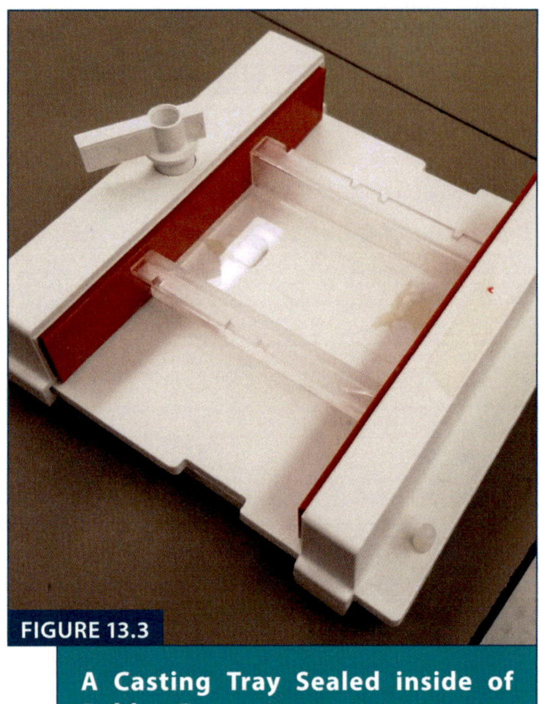

FIGURE 13.3

A Casting Tray Sealed inside of Rubber Dams

8. Allow the gel to solidify (approximately 30 minutes).

Thinking Critically: Restriction Enzymes

Restriction enzymes are often derived from bacterial cells, which use them as a form of defense. Knowing what you do about restriction enzymes, work with your lab partners to come up with an explanation as to what a restriction enzyme could protect a bacterial cell against.

PROCEDURE 13.2: LOADING DNA SAMPLES

1. Once your gel has cooled, remove the comb from the gel plate. Notice that the comb has left indentations (called wells) in the plate. You will load DNA samples into these wells.

2. Next, locate your electrophoresis chamber (see Figure 13.4) and remove the top of the chamber. Notice that the chamber has connection ports for a black (negative) and red (positive) plug. When fully assembled, the device will run an electric current through the gel plate. DNA is negatively charged, which will cause it to move toward the positive end of the apparatus (red port).

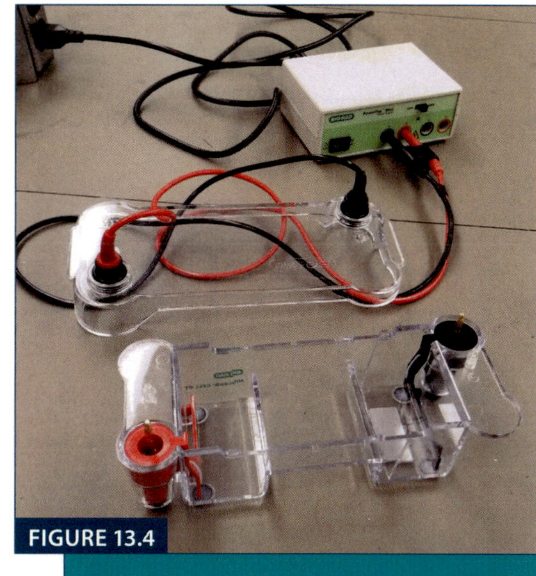

FIGURE 13.4

An Electrophoresis Chamber

3. Gently loosen the rubber dams and remove your casting tray (with the gel plate still inside of the tray). Place the casting tray and gel inside your electrophoresis chamber bottom, *making sure that the wells face the black (negative) port.*

4. Fill your large beaker with 350 mL of 1× electrophoresis buffer from the carboy and pour the solution into the bottom half of your chamber. The fluid should cover the surface of your gel plate.

Before you begin loading your DNA samples, your instructor will explain how to use a micropipette (see Figure 13.5). Set your micropipette to 20 microliters (μL), and locate the *practice dye* in your microcentrifuge tube rack. Your instructor will give you a practice gel plate (made of rubber) submerged in water to practice loading 20 μL of dye into each well. This requires precision, as your instructor will demonstrate. As you practice, try to get a feel for the sides of the wells with your micropipette tip as you prepare to eject the dye.

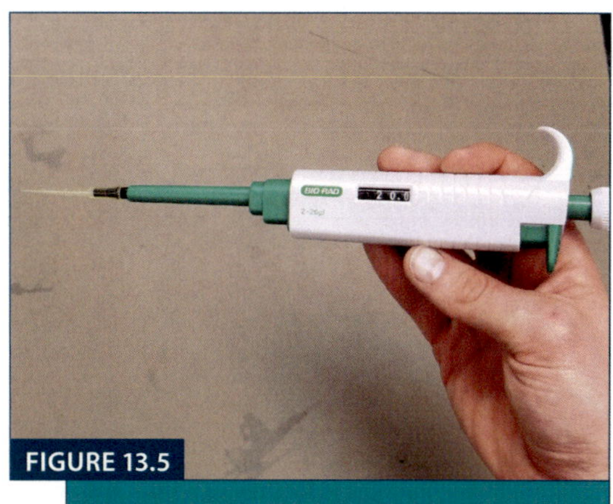

FIGURE 13.5

A Micropipette with Tip Attached

Once you are done practicing, you are ready to load your DNA samples (labeled components A–F).

1. Beginning with the well in the upper left corner of your plate, load 20 μL from component A into the well. Be careful not to poke through the bottom of your gel plate!

2. Eject your used micropipette tip, and follow the same procedure for components B–F in the remaining wells.

3. Once you have loaded all of the DNA samples, attach the top of the electrophoresis chamber (matching the red and black components), plug it into the power source, and switch the button on the power source to the "on" position. You should begin to see bubbles forming in your chamber.

4. Allow the samples to run for about 40–45 minutes. As you wait, begin washing out your practice plate and throwing away your used micropipette tips.

5. After your electrophoresis experiment is complete, turn off the power source. Wait a minute or two until the device cools before removing your casting tray.

6. Remove the casting tray from the chamber, but *be very careful* to gently hold the gel plate in place with your index finger, as it is slippery and can slide off the tray as you transfer the gel. Allow the gel to slide off your tray and into the container of staining dye provided by your instructor. The plates will stain until the next lab period.

PROCEDURE 13.3: ANALYSIS OF ELECTROPHORESIS RESULTS

Locate your plate from the staining tray (your instructor may pass them out before class) and place your plate on top of one of the lamps on the side bench to illuminate the plate from the bottom. Observe the patterns formed by the DNA fragments that filtered through the gel plate. Remember that the samples you loaded were as follows:

Component A: Crime Scene DNA treated with Restriction Enzyme 1

Component B: Crime Scene DNA treated with Restriction Enzyme 2

Component C: First Suspect's DNA treated with Restriction Enzyme 1

Component D: First Suspect's DNA treated with Restriction Enzyme 2

Component E: Second Suspect's DNA treated with Restriction Enzyme 1

Component F: Second Suspect's DNA treated with Restriction Enzyme 2

This gave you fragment patterns in six different lanes (where the DNA traveled from each well). The first two lanes contain the fragments from the crime scene DNA, the second two lanes contain filtered DNA from the first suspect, and the last two lanes contain filtered DNA from the second suspect. The reason the samples are in pairs is because each DNA sample was treated with *two* different restriction enzymes. Draw your electrophoresis plate below, including where (approximately) the bands formed in each lane.

Based on the patterns of the DNA fragments produced through electrophoresis (the RFLPs), which suspect's DNA matches that of the crime scene? ________________

Could these DNA samples have been distinguished from one another if *only* enzyme 1 were used? Explain why or why not below.

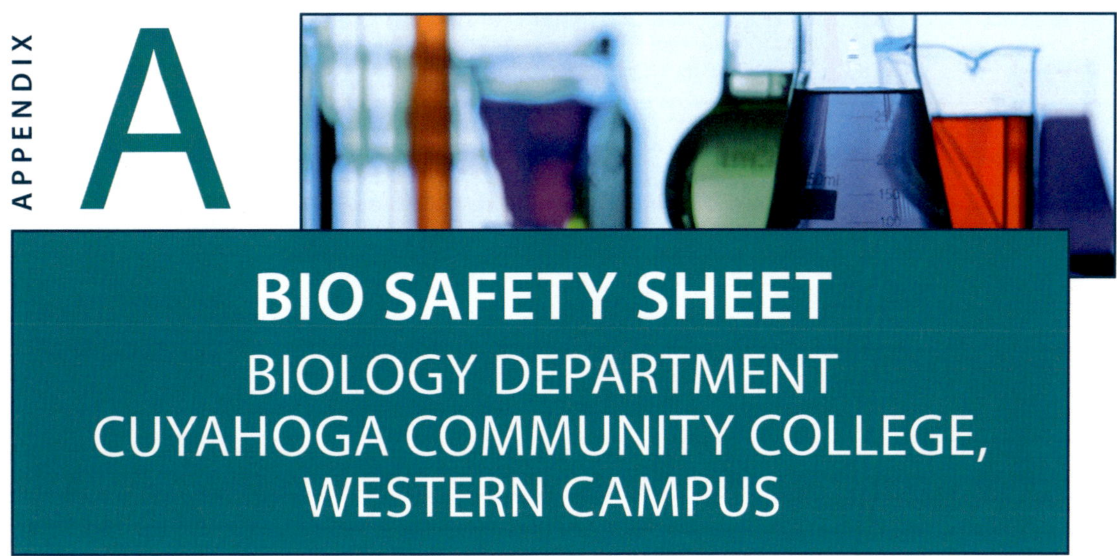

BIO SAFETY SHEET
BIOLOGY DEPARTMENT
CUYAHOGA COMMUNITY COLLEGE, WESTERN CAMPUS

GENERAL LABORATORY INSTRUCTIONS

1. Avoid waste of gas, water (both tap and distilled), and materials of any kind.

2. Use only your reagents and return reagent bottles promptly to their place. (Do not return excess reagents to stock bottles.)

3. Dispose of hazardous chemicals by placing them in hazardous waste jars provided for this purpose.

4. Notify your instructor of any chemical spills immediately. Clean up chemical spill from floors and countertops immediately before other people can come in contact with them.

5. Before leaving the laboratory make sure that the gas and water are turned off, your desk is clean and your chair is pushed in.

6. Maintain an orderly arrangement of the equipment in your community drawer and/or tote trays.

7. Use only the equipment and supplies assigned to you and only after your instructor has explained their operation to you.

8. Never walk away from a lit Bunsen burner.

SPECIAL SAFETY RULES

1. Wear approved eye protection at all times in the laboratory when chemicals are in use. Approved eye wear protects against both impact and splashes. (If you should get a chemical in your eye, flush with flowing water from the eye wash station.)

2. Perform no unauthorized experiments.

3. In case of fire or accident, call your instructor immediately. (Note the location of the fire extinguisher, fire blanket, spill tamer, and safety shower now so you can access them quickly if necessary.)

4. Do not taste anything in the laboratory. Do not eat or drink in the laboratory or use lab glassware for this purpose.

5. Use great care in noting the odor of any fumes and avoid breathing in chemical fumes of any kind.

6. Do not use mouth suction to fill pipettes with chemical reagents. Always use a pipette pump or bulb.

7. Protect your hands with paper towels when putting glass tubing into rubber stoppers.

8. Tie back your hair when working in the laboratory. A lab apron or lab coat is essential when wearing combustible clothing and provides desirable protection on all occasions.

9. Never work in the lab without an instructor present.

B

DNA EXTRACTION AT HOME

To conduct your own DNA extraction at home, you will need the following materials:

- About 1 half cup of your DNA source (any type of fruit or vegetable). Frozen fruits tend to work best.

- ⅛ teaspoon of salt

- A strainer (I use the metal kind). If you don't have one, you may borrow some gauze from lab.

- Dish soap

- Isopropyl (rubbing) alcohol. This can be either 70 or 90%. Ethyl alcohol works too. Put your bottle in the freezer for optimal extraction!

- Meat tenderizer OR pineapple juice (either contains an enzyme that helps separate the DNA)

- A small drinking glass or glass tube

PROCEDURE

1. Place your half cup of fruit/vegetable in a blender with 1 cup of cold water. Add the ⅛ teaspoon of salt and blend for about 15 seconds or until you get a soupy mixture of one solid color.

2. Pour your mixture through a strainer into a container (i.e., a measuring cup).

3. Add 2 tablespoons of dish soap into the container and *gently* stir.

4. Let the mixture sit for 5–10 minutes.

5. Add a pinch of meat tenderizer or 1 mL of pineapple juice to the container and *gently* stir.

6. Pour the mixture into a small glass container (i.e., a test tube) until it is about half full. Since you might not have a tube, you can use a small glass (I used a shot glass).

7. Fill the glass container with cold iso-propyl alcohol until it is almost full. You should see two layers formed. The alcohol will sit on top of the blended food mixture. Between the two layers should be a cloudy/bubbly film, which is DNA.

FIGURE B.1

DNA Extraction